E. Marchionna (Ed.)

Questions on Algebraic Varieties

Lectures given at a Summer School of the
Centro Internazionale Matematico Estivo (C.I.M.E.),
held in Varenna (Como), Italy,
September 7-17, 1969

Springer

FONDAZIONE
CIME
ROBERTO CONTI

C.I.M.E. Foundation
c/o Dipartimento di Matematica "U. Dini"
Viale margagni n. 67/a
50134 Firenze
Italy
cime@math.unifi.it

ISBN 978-3-642-11014-6 e-ISBN: 978-3-642-11015-3
DOI:10.1007/978-3-642-11015-3
Springer Heidelberg Dordrecht London New York

Reprint of the 1st ed. C.I.M.E., Ed. Cremonese, Roma 1970
With kind permission of C.I.M.E.

Printed on acid-free paper

Springer.com

CENTRO INTERNAZIONALE MATEMATICO ESTIVO
(C. I. M. E.)

3° Ciclo - Varenna - dal 7 al 17 Settembre 1969

« QUESTIONS ON ALGEBRAIC VARIETIES »

Coordinatore: Prof. E. MARCHIONNA

CENTRO INTERNAZIONALE MATEMATICO ESTIVO

(C.I.M.E.)

Pierre DOLBEAULT

RESIDUS ET COURANTS

Corso tenuto a Varenna dal 7 al 17 Settembre 1969

RESIDUS ET COURANTS

par

Pierre DOLBEAULT
(Université de Poitiers)

1. Introduction:

Soit X une surface de Riemann et soit g une fonction méromorphe sur un voisinage U d'un point P de X ayant pour seul pôle P ; soit z une coordonnée locale sur U telle que $z(P) = 0$, alors g est holomorphe sur $U \setminus P$ et égale au voisinage de P à une série de Laurent dont le coefficient du terme en z^{-1} est appelé le résidu α de g en P ; en fait α est un invariant de la forme différentielle méromorphe fermée $\omega = g(z)dz$.

Sur $U \setminus P$, la forme différentielle ω définit un courant $\underline{\omega}$, c'est-à-dire une forme linéaire continue sur l'espace $\mathcal{D}(U \setminus P)$ des formes différentielles φ , de classe C^∞ , à support compact dans $U \setminus P$, par la formule :

$$\underline{\omega}\,[\varphi] = \int_U' \omega \wedge \varphi \, .$$

Considérons maintenant, pour toute $\psi \in \mathcal{D}(U)$, la limite :

$$T[\psi] = \lim_{\varepsilon \to 0} \int_{|z| \geq \varepsilon} \omega \wedge \psi \quad \text{(valeur principale de Cauchy)}$$

Il est facile de vérifier que cette limite existe et définit un courant dont la restriction à $U \setminus P$ est $\underline{\omega}$. De plus, si $d = d' + d''$ est la différentiation extérieure, d'' étant la partie de d qui augmente le degré en $d\bar{z}$, on a :

$$d''T = 2\pi i \alpha \, \delta_P + d'B \, ,$$

où δ_P est la mesure de Dirac de support P et B un courant. En particulier, on a : $B = 0$ si P est un pôle simple.

Soit maintenant X une variété analytique complexe paracompacte, de dimension complexe n. On dit qu'une forme différentielle ω de degré p définie sur X est méromorphe si elle contient seulement des différentielles des coordonnées locales complexes et si ses coefficients sont des fonctions méromorphes ; ω est dite semi-méromorphe si, localement, c'est le produit d'une forme différentielle α, de classe C^∞ par $\frac{1}{f}$ où f est une fonction holomorphe; il est clair que l'on peut définir le faisceau des germes de formes différentielles semi-méromorphes et qu'une forme semi-méromorphe sur un ouvert U de X est une section de ce faisceau au-dessus de U. Si $\omega = \frac{\alpha}{f}$ sur l'ouvert U, on appelle l'ensemble $S = \{x \in U \mid f(x) = 0\}$ un ensemble polaire de ω sur U ; c'est un ensemble analytique de U de codimension complexe 1 dans le support de α ; il contient l'ensemble des points où ω n'est pas C^∞ ; le plus petit de ces ensembles sera dit l'ensemble polaire de ω / U.

Supposons maintenant $d\omega = o$; (ω fermée). A la suite de H. Poincaré (1887) [8], J. Leray (1959) [7] a défini, dans le cas des pôles simples et lorsque S est lisse (sans points singuliers) une $(p-1)$-forme différentielle fermée sur S qui généralise le nombre résidue α en P ci-dessus. Avant lui, également guidés par le mémoire de H. Poincaré, Kodaira (1951; 1952) [5], [6], L. Schwartz (1953) [9] et P. Dolbeault (1957) [1] ont défini des courants généralisant le courant résidu $\alpha\delta_P$ pour des singularités particulières de S

Voici un essai de définition d'un courant résidu quelles que soient les singularités de S par utilisation de résultats connus sur les résolutions des singularités.

Un opérateur différentiel D sur l'espace $\mathcal{D}'(X)$ des courants sur X est dit semi-holomorphe si, pour tout $x \in X$, il existe un voisinage U de x sur lequel des coordonnées $(z_1, \ldots, z_n)$ sont définies et tel que, sur U : $D = \sum_{i_1, \ldots, i_n} \alpha_{i_1 \ldots i_n}(z) \frac{\partial^{i_1}}{\partial z_1^{i_1}} \cdots \frac{\partial^{i_n}}{\partial z_n^{i_n}}$, où les $\alpha_{i_1 \ldots i_n}$ sont des fonctions C^∞.

Cette définition est indépendante du système de coordonnées et D opère également sur les formes semi-méromorphes [9]

Un problème préliminaire est le suivant : Soit ω une forme différentielle semi-méromorphe sur X, d'ensemble polaire S, construire un courant $T(\omega)$ prolongeant le courant $\underline{\omega}$ défini sur $X - S$ par ω et tel que, pour tout opérateur semi-holomorphe D, on ait : $T(D\omega) = DT(\omega)$.

Nous donnons d'abord une solution dans le cas de singularités assez simples (cas normal) puis une construction de $T(\omega)$ par réduction de singularités (Hironaka [3]) lorsque X est une variété algébrique, S une sous-variété algébrique de X. Pour terminer, nous appliquons le résultat à des exemples de résidus.

2. Le cas normal :

C'est une généralisation du cas régulier introduit par L. Schwartz [9].

2.1. Une forme différentielle semi-meromorphe ω_x définie au voisinage de $x \in X$ est dite élémentaire s'il existe un système de coordonnées $(z_1, \ldots, z_n)$ en x tel qu'un ensemble polaire de ω soit la réunion des ensembles $z_1 = o$, $z_2 = o, \ldots,$ $z_p = o$ $(p \leq n)$.

Toute forme semi-méromorphe ω sur X qui est, au voisinage de tout point de $x \in X$, égale à une somme finie de formes élémentaires est dite normale.

L'ensemble des opérateurs différentiels semi-holomorphes constitue un anneau $\triangle$ la multiplication étant définie par la composition des opérateurs.

2.2. Lemme : L'espace **C**-vectoriel des formes différentielles semiméromorphes normales est un $\triangle$-module engendré, sur tout compact, par les formes à coefficients localement sommables.

Démonstration : Soit ω une forme semi-méromorphe normale sur X ; pour tout $x \in X$, il existe un voisinage ouvert U_x sur lequel ω est une somme finie de formes élémentaires ; soit $(u_i)_{i \in I}$ un sous-recouvrement localement fini de $(U_x)_{x \in X}$. Soit $\sum_{i \in I} \gamma_i$ une partition de l'unité subordonnée à $(u_i)_{i \in I}$; c'est un élément de $\triangle$ et ω est égale à la somme localement finie $\sum_{i \in I} \omega_i$, avec $\omega_i = \gamma_i \omega$. La forme ω_i est une somme finie de formes élementaires à support compact. Donc, sur tout compact K, ω est égale à une somme finie de formes élémentaires à support compact dans X, chaque support étant contenu dans un u_i ; il suffit d'établir que toute forme élémentaire, à support compact dans u_i, est l'image, par un élément de $\triangle$, d'une forme différentielle à coefficients localement sommables. Soit ω' une telle forme élémentaire, il existe un système de coordonnées $(z_1, \ldots, z_n)$ défini au voisinage de son support et une forme C^∞ α tels que :

$$\omega' = \frac{\alpha}{z_1^{p_1} \dots z_n^{p_n}}, \text{ avec } p_k \geqslant 0 \text{ pour } k = 1, \dots, n .$$

Si $p_s \geqslant 2$, on a : $\omega' = -\frac{1}{p_s - 1} \frac{\partial}{\partial z_s} \left(\frac{\alpha}{z_1^{p_1} \dots z_s^{p_s - 1} \dots z_n^{p_n}} \right) + \frac{1}{p_s - 1} \frac{(\partial \alpha / \partial z_s)}{z_1^{p_1} \dots z_s^{p_s - 1} \dots z_n^{p_n}}$

Par récurrence,on obtient : $\omega' = \sum_\mu D_\mu \, \omega''_\mu$ où D_μ est un opérateur

différentiel semi-holomorphe et $\omega''_\mu = \dfrac{\alpha''_\mu}{z_{i_1} \dots z_{i_p}}$ $(1 \leqslant i_1 < \dots < i_p \leqslant n)$;

ω''_μ est à coefficients localement sommables.

2.3 <u>Théorème</u> : <u>Pour toute forme différentielle semi-méromorphe normale</u> ω , <u>il existe un courant unique</u> $T(\omega)$ <u>tel que</u> :

(1) <u>si</u> ω <u>a des coefficients localement sommables, alors</u> $T(\omega)$ <u>coincide avec le courant</u> $\underline{\omega}$ <u>défini par</u> : $\underline{\omega}[\varphi] = \int_X \omega \wedge \varphi$;

(2) <u>l'opérateur</u> $T : \omega \to T(\omega)$ <u>est linéaire pour les structures de</u> Δ<u>-modules de l'espace</u> $\mathcal{S}$ <u>des formes différentielles semi-méromorphes normales et de l'espace des courants, en particulier pour tout</u> $D \in \Delta$, <u>pour toute</u> $\omega \in \mathcal{S}$, <u>on a</u> : $T(D\,\omega) = DT(\omega)$;

(3) <u>l'opérateur</u> T <u>est local</u> (i.e. : <u>si</u> U <u>est un ouvert de</u> X <u>alors</u> : $T(\omega | U) = T(\omega) | U$) , <u>et supp</u> $T(\omega) = \text{supp } \omega$.

En particulier, le <u>courant</u> $T(\omega)$ <u>prolonge canoniquement</u> ω <u>de</u> $X \setminus S$ <u>à</u> X <u>et satisfait à</u> (2) . La démonstration du théorème sera faite de 2,4. à 2.6 .

2.4. Lemme : Pour toute forme ω élémentaire sur un ouvert, il existe des formes ω_μ, à coefficients localement sommables et des opérateurs semiholomorphes D_μ tels que

$$\omega = \sum_\mu D_\mu \, \omega_\mu \tag{4}$$

et $\operatorname{supp}(\omega_\mu) \subset \operatorname{supp}(\omega)$.

Démonstration : Cela résulte de la construction des formes ω_μ donnée dans la démonstration du lemme 2.2.

2.5. Lemme : S'il existe un opérateur T satisfaisant à (1) et à (2) de 2.3, alors T est unique et satisfait à (3) de 2.3. De plus pour établir l'existence de $T(\omega)$, il suffit de le faire dans le cas où ω est élémentaire.

Démonstration : D'après 2.2, il existe des formes ω_ν à coefficients localement sommables et des opérateurs semi-holomorphes D_ν tels que : (5) $\omega = \sum_\nu D_\nu \, \omega_\nu$.

Alors (6) $T(\omega) = T(\sum_\nu D_\nu \omega_\nu) = \sum_\nu D_\nu \, T(\omega_\nu) = \sum_\nu D_\nu \, \omega_\nu$;

les deux dernières égalités résultent de (2) et (1) respectivement

Cela prouve que l'existence de $T(\omega)$ entraîne son unicité.

Pour tout ouvert U de X, on a:

$$\omega | U = (\sum_\nu D_\nu \, \omega_\nu) | U = \sum_\nu ((D_\nu \, \omega_\nu) | U) = \sum_\nu D_\nu(\omega_\nu | U)$$

parce que les D_ν sont des opérateurs locaux.

Alors : $T(\omega) | U) = \sum_\nu D_\nu \, \underline{(\omega_\nu | U)} = \sum_\nu D_\nu(\underline{\omega}_\nu | U) = T(\omega) | U$, ce qui prouve le caractère local de T.

Soit U un ouvert de X sur lequel : $\omega = \omega_1 + \ldots + \omega_p$, où $\omega_k (k = 1, \ldots, p)$ est élémentaire sur U . Soit $x \notin \mathrm{supp}(\omega)$, alors il existe un voisinage V de x sur lequel : $\omega | V = 0$; on désignera $\omega | V$ par ω et $\omega_k | V$ par ω_k pour simplifier la notation; on a :

(7) $$\omega_1 = -(\omega_2 + \ldots + \omega_p) = \omega'' .$$

Il existe un système de coordonnées $(z_1, \ldots, z_n)$ sur V dans lequel:

$\omega_1 = \dfrac{\alpha'}{z_1^{p_1} \ldots z_n^{p_n}}$ où α' est C^∞ ; alors :

ou bien l'ensemble polaire de ω_1 est contenu dans celui de ω'' ; ou bien il existe j ($1 \leqslant j \leqslant n$) , par exemple $j = 1$ après changement éventuel du numérotage des coordonnées, tel que $\{z_1 = 0\}$ ne soit pas une composante de l'ensemble polaire de ω'' . Multiplions les deux membres de (7) par $z_2^{p_2} \ldots z_n^{p_n}$; alors :

$z_1^{-p_1} \alpha' = z_2^{p_2} \ldots z_n^{p_n} \omega''$. Mais $z_1^{-p_1} \alpha'$ est C^∞ sur le complémentaire du sous-ensemble rare et fermé de $\{z_1 = o\} \cap V$, intersection de l'ensemble polaire de ω'' et de $z_1 = o$; donc $z_1^{-p_1} \alpha'$ est C^∞ sur V d'après ([2] , th. 1; [2bis] , 4.2) .

Par récurrence, on est ramené au premier cas.

$\omega'' = \dfrac{\beta}{f_1^{q_1} \ldots f_m^{q_m}}$ où chacune des f_j est une fonction coordonnée sur V , mais où on peut avoir $m > n$; si $f_1 = o$ fait partie de l'ensemble polaire de ω_1 , au produit près par une fonction holomorphe sans zéro , il existe k tel que z_k soit égal à f_1 ; sinons, considérons :

$$f_1^{-q_1} \beta_1 = f_2^{q_2} \ldots f_m^{q_m} \omega'' \quad ;$$

cette forme est C^∞ sur le complémentaire du sous-ensemble rare et fermé de $f_1 = o$ intersection de $f_1 = o$ et de l'ensemble polaire de ω_1 ; donc est C^∞ sur V , d'après ([2] , th. 1 ; [2bis] , 4.2) .

Par récurrence sur l'indice l de f_l , $(l = 1, \ldots, m)$, on montre ainsi que l'ensemble polaire de ω'' est contenu dans celui de ω_1 , donc ω'' est une forme élémentaire.

Alors $\omega = \omega_1 - \omega''$ est une forme élémentaire sur V , nulle sur V ; en vertu du lemme 2.4, dans l'expression (4) de ω , les formes ω_ν sont nulles, donc si $T(\omega)$ est défini pour toute forme élémentaire, on a $T(\omega) = o$, en vertu de l'expression (6) Autrement dit, pour toute forme ω , $T(\omega)$ est indépendant de la décomposition (5) , donc aussi de la décomposition en formes élémentaires $\omega = \omega_1 + \ldots + \omega_p$. Pour établir l'existence de $T(\omega)$ il reste à établir l'existence de $T(\omega')$ pour toute forme élémentaire ω' et à poser $T(\omega) = \sum_{k=1}^{p} T(\omega_k)$

D'autre part, on a supp $T(\omega) \supset$ supp ω à cause de (1) et du fait que tout ensemble polaire de ω est contenu dans l'adhérence de l'ensemble des points où ω est C^∞ ; d'autre part si T existe et satisfait à (1) et (2) , alors d'après le résultat ci-dessus supp $T(\omega) \subset$ supp ω , d'où la propriété du support énoncée dans (3) .

2.6 . Existence de $T(\omega)$ satisfaisant à (1) est (2) pour ω élémentaire.

a) Soit $\omega = \dfrac{\alpha}{z_1^{p_1} \dots z_n^{p_n}}$ où $z_1, \dots, z_n$ sont des coordonnées locales complexes dans un ouvert U de X et α une forme différentielle C^∞ ; pour toute $\varphi \in \mathcal{D}(U)$, soit $\alpha \wedge \varphi = \psi \, dz_1 \wedge d\bar{z}_1 \wedge \dots \wedge dz_n \wedge d\bar{z}_n$; posons :

$$T(\omega)[\varphi] = \lim_{\varepsilon_n \to 0} \int_{|z_n| \geq \varepsilon_n} \frac{dz_n \wedge d\bar{z}_n}{z_n^{p_n}} \left(\lim_{\varepsilon_{n-1} \to 0} \int_{|z_{n-1}| \geq \varepsilon_{n-1}} \frac{dz_{n-1} \wedge d\bar{z}_{n-1}}{z_{n-1}^{p_{n-1}}} (\dots \right.$$

$$\left. \dots \left(\lim_{\varepsilon_1 \to 0} \int_{|z_1| \geq \varepsilon_1} \frac{\psi \, dz_1 \wedge d\bar{z}_1}{z_1^{p_1}} \right) \dots \right) .$$

On peut prouver, à l'aide de la formule de Taylor en $z_1, \bar{z}_n$ que :

$$\lim_{\varepsilon \to 0} \int_{|z_1| \geq \varepsilon} \frac{\psi(z_1, \dots, z_n)}{z_1^{p_1}} dz_1 \wedge d\bar{z}_1$$

est une fonction C^∞ des parties réelles et imaginaires de $z_2, \dots, z_n$. On définit $T(\omega)[\varphi]$ par récurrence sur n .

φ étant à support compact, on peut intervertir les passages à la limite et les intégrations, de sorte que :

$$T(\omega)[\varphi] = \lim_{\varepsilon_n \to 0} \Big(\lim_{\varepsilon_{n-1} \to 0} \dots \Big(\lim_{\varepsilon_1 \to 0} \int_{|z_n| \geq \varepsilon_n} \Big(\int_{|z_{n-1}| \geq \varepsilon_{n-1}} \dots$$

$$\dots \Big(\int_{|z_1| \geq \varepsilon_1} \frac{\psi \, dz_1 \wedge d\bar{z}_1 \wedge \dots \wedge dz_n \wedge d\bar{z}_n}{z_1^{p_1} \dots z_n^{p_n}} \Big) \dots \Big) .$$

On vérifie aisément que $T(\omega)$ est un courant .

b) Soit D un opérateur différentiel semi-holomorphe, il s'agit de démontrer : $T(D\,\omega) = DT(\omega)$. Il suffit de considérer les deux cas : D est la multiplication par une fonction $C^\infty \beta$ et $D = \frac{\partial}{\partial z_k}$.

On a évidemment : $T(\omega)[\beta\varphi] = T(\beta\omega)[\varphi]$. Reste à prouver :

$$- T(\omega)\left[\frac{\partial \psi}{\partial z_k}\right] = T\left(\frac{\partial \omega}{\partial z_k}\right)[\psi] .$$

Désignons par ω_J et α_L les coefficients de ω et φ, J et L étant des multi-indices convenables.

On peut commencer l'intégration par z_k et passer à la limite pour $\varepsilon_k \to 0$, sans changer $T(\omega)[\psi]$; mais :

$$\omega \wedge \psi = \sum_{JL} (-1)^I \omega_J \, \alpha_L \, dz_1 \wedge d\bar{z}_1 \wedge \dots \wedge dz_n \wedge d\bar{z}_n ,$$ où J, L sont des multi-indices dont la réunion est celle des indices de $z_1, \bar{z}_1, . \, z_n, \bar{z}_n$ et I la signature de la permutation faisant passer de J, L à l'ensemble ordonné des indices des z_1 et des $\bar{z}_1$; de plus :

$$\omega_J = \frac{\alpha_J}{z_1^{p_1} \ldots z_n^{p_n}} \int_{|z_k| \geqslant \varepsilon} \frac{\alpha_J}{z_k^{p_k}} \frac{\partial \psi_L}{\partial z_k} dz_k \wedge d\bar{z}_k + \int_{|z_k| \geqslant \varepsilon} \frac{\partial}{\partial z_k} \left(\frac{\alpha_J}{z_k^{p_k}}\right) \Psi_L dz_k \wedge d\bar{z}_k$$

$$= \int_{|z_k| \geqslant} \frac{\partial}{\partial z_k} \left(\frac{\alpha_J \psi_L}{z_k^{p_k}}\right) dz_k \wedge d\bar{z}_k = \int_{|z_k| \geqslant \varepsilon} d\left(\frac{\alpha_J \psi_L}{z_k^{p_k}} d\bar{z}_k\right) =$$

$$= - \int_{|z_k| = \varepsilon} \frac{\alpha_J \Psi_L}{z_k^{p_k}} d\bar{z}_k \quad \text{qui tend vers } 0$$

quand ε tend vers 0 .

c) Il est clair que, si ω a des coefficients localement sommables $T(\omega)$ coincide avec $\underline{\omega}$.

3. Réduction au cas normal :

3 . 1 Nous allons considérer le cas particulier suivant qui peut, probablement, être généralisé au cas de n'importe quelle variété analytique complexe (travaux non publiés d'Hironaka) .

Soit X une variété algébrique projective irréductible lisse de dimension n sur $\mathbb{C}$; c'est une variété analytique complexe. Soit ω une p-forme différentielle semi-méromorphe sur X dont l'ensemble polaire est contenu dans une sous-variété algébrique S de X , de codimension 1 . On sait (conjecture de Hodge Atiyah ([4] , p. 81) prouvée par Hironaka ([3] , p. 146)) qu'il existe une transformation birationnelle $\pi : X' \longrightarrow X$ d'une variété algébrique lisse X' sur X, qui est un morphisme et telle que $S' = \pi^{-1}(S)$ soit une sous-variété de X' de codimension 1, égale aù voisinage de tout point x' à la réunion des ensembles analytiques : $z_1 = 0$; ...; $z_p = 0$ où $(z_1, \dots, z_n)$ est un système de coordonnées locales au voisinage de x' et où $p \leqslant n$.

π est un morphisme analytique, donc de classe C^∞ et est propre; de plus $\pi \mid X' \setminus S'$ est un isomorphisme analytique : $X' \setminus S' \rightarrow X \setminus S$. On dira que π est un morphisme normalisant pour S.

3.2. Image réciproque de ω. Soit $x \in X$; il existe un voisinage U de x sur lequel ω est égale à : $f^{-1}\alpha$ où α est une forme C^∞ et f une fonction holomorphe sur U telle que : $S \cap U = \{y \in U \mid f(y) = 0\}$; on a : $f\omega =$ et, en dehors de S, $\pi^*(f\omega) = \pi^* f \,.\, \pi^* \omega = \pi^* \alpha$; on pose, sur U : $\pi^* \omega = \dfrac{\pi^*}{\pi^* f} \,.\, \pi^* \omega$ est semi-méromorphe sur $\pi^{-1}(U)$. L'ensemble polaire de $\pi^* \omega$ est contenu dans S' puisque :

$$\{y' \in \pi^{-1}(U) \mid \pi^* f(y') = 0\} = \{y' \in \pi^{-1}(U) \mid f \circ \pi(y') = 0\} = \pi^{-1}(U) \cap \pi^{-1}(S) .$$

La forme $\omega' = \pi^* \omega$ définie comme ci-dessus au voisinage de tout point de X' est alors une forme différentielle semi-méromorphe normale sur X' ; en fait, au voisinage de tout point de X', elle est égale à une forme élementaire.

3.3. Soit $T' = T'(\omega')$ le prolongement à X' du courant $\underline{\omega}'$ sur $X' \setminus S'$ défini en 2.3. L'application π étant propre, le courant $\pi_* T'$ est défini sur X de la façon suivante: pour toute $\varphi \in \mathcal{D}(X)$, on a : $\pi^*\varphi \in \mathcal{D}(X')$, alors : $\pi_* T'[\varphi] = T'[\pi^*\varphi]$.

Posons: $T_\pi(\omega) = \pi_* T'$; ce courant est complètement déterminé par ω et ; il possède les propriétés qui seront énumérées de 3.4. à 3.8 .

3.4. Lemme : Si ω est à coefficients localement sommables, alors $T_\pi(\omega) = \underline{\omega}$.

Démonstration : Soit $\varphi \in \mathcal{D}(X)$, alors, parce que S est de mesure nulle sur X, que $\pi | X' \setminus S'$ est un isomorphisme, et d'après la construction explicite de $T'(\pi^*\omega)$, on a :

$$\underline{\omega}[\varphi] = \int_X \omega \wedge \varphi = \int_{X \setminus S} \omega \wedge \varphi = \int_{X' \setminus S'} \pi^*\omega \wedge \pi^*\varphi$$

$$= T'(\pi^*\omega)[\pi^*\varphi] = T_\pi(\omega)[\varphi].$$

3.5 Lemme : T_π est local et supp $T_\pi(\omega) = \mathrm{supp}(\omega)$. Cela résulte de 2.3 (3) appliqué à $T'(\omega')$.

3.6. Lemme : Pour tout opérateur différentiel semi-holomorphe D , on a : $T_\pi(D\omega) = D\, T_\pi(\omega)$.

Démonstration : Pour toute forme semi-méromorphe ω , pout tout opérateur semi-holomorphe D , l'ensemble polaire de $D\omega$ est contenu dans celui de ω . Pour toutes les formes semi-méromorphes dont l'ensemble polaire est contenu dans le même ensemble algébrique

S , on peut utiliser le même morphisme π .

Il suffit de montrer que, pour tout domaine de carte U dans X , on a, sur U :

$$\frac{\partial}{\partial z_1} T_\pi(\omega) = T_\pi\left(\frac{\partial}{\partial z_1}\omega\right) .$$

et que , pour toute fonction C^∞ α sur U , $\alpha\, T_\pi(\omega) = T_\pi(\alpha\,\omega)$.

(a) Nous supposons, dans la suite, que U est relativement compact dans X (ce qui est possible puisque X est localement compact), alors $U' = \pi^{-1}(U)$ est relativement compact dans X' , donc est recouvert par un nombre fini de domaines de cartes $U'_1, \dots, U'_L$ relativement compacts suffisamment petits pour que, dans chacun d'eux, $\pi^*\omega$ soit élémentaire.

On considère la restriction de ω à U que l'on note encore ω . Par définition :

$$T_\pi\left(\frac{\partial\omega}{\partial z_1}\right)[\varphi] = T'\left(\pi^*\frac{\partial\omega}{\partial z_1}\right)[\pi^*\varphi] .$$

Soit $\sum_{i=1}^{L}\psi'_i$ une partition de l'unité sur $\pi^{-1}(U)$ subordonnée au recouvrement $(U'_1, \dots, U'_L)$ de U', alors :

$$T'\left(\pi^*\frac{\partial\omega}{\partial z_1}\right)[\pi^*\varphi] = \sum_{i=1}^{L} T'\left(\pi^*\frac{\partial\omega}{\partial z_1}\right)[\psi'_i\,\pi^*\varphi]$$

Soient $(z'^i_1, \dots, z'^i_n)$ des coordonnées sur U'_i telle que $S' \cap U'_i$ soit contenu dans l'ensemble analytique $\bigcup_{\ell=1}^{n}\{z'^i_\ell = 0\}$.

Considérons $\psi'_i\, T'\,(\pi^* \frac{\partial\omega}{\partial z_1})[\pi^*\varphi]$

$$= T'(\pi^* \frac{\partial\omega}{\partial z_1})[\psi'_i \pi\varphi | U'_i] ,$$

parce que, dans la dernière expression, le support du courant considéré est contenu dans U'_i .

Posons : $\lim\limits_{\varepsilon' \to o} \int_{U'_{i\,\varepsilon'}} = \lim\limits_{\varepsilon'_n \to o} (\ldots \lim\limits_{\varepsilon'_1 \to o} \int_{U'_i\, |z'^i_n| \geq \varepsilon'_n \ldots |z'^i_1| \geq \varepsilon'_1}) \ldots) ,$

alors par définition de T' :

$$(8) \qquad T'\,(\pi^* \frac{\partial\omega}{\partial z_1}\left[\psi'_i \pi^*\varphi | U'_i\right] = \lim_{\varepsilon' \to o} \int_{U'_{i\,\varepsilon'}} \pi^* \frac{\partial\omega}{\partial z_1} \wedge \psi'_i\, (\pi^*\varphi | U'_i) .$$

Considérons $\pi_i = \pi | U'_i$; $\pi_i \mid U'_i \setminus S'$ est un isomorphisme de $U'_i \setminus S'$ sur $\pi(U'_i \setminus S') \subset U \setminus S$; sur $\pi_i(U'_i \setminus S')$, posons $\psi_i = (\pi_i^{-1})^* \psi'_i$; alors : $\psi'_i = \pi_i^* \psi_i$ sur $U'_i \setminus S'$.

L'expression (8) est égale à :

$$\lim_{\varepsilon' \to o} \int_{U'_{i\,\varepsilon'}} \pi_i^* \frac{\partial\omega}{\partial z_1} \wedge \pi_i^*(\psi_i \varphi) = \lim_{\varepsilon' \to o} \int_{U'_{i\,\varepsilon'}} \pi_i^*\; \omega \wedge \pi_i^* \frac{\partial}{\partial z_1} (\psi_i \varphi) +$$

$$+ \lim_{\varepsilon' \to o} \int_{U'_{i\,\varepsilon'}} \pi_i^* \frac{\partial}{\partial z_1} (\omega \wedge \psi_i \varphi) .$$

Posons : $\omega \wedge \psi_i \varphi = u\, d\, z_1 \wedge d\, \bar{z}_1 \wedge \ldots \wedge d\, z_n \wedge d\, \bar{z}_n$, alors :

$$\frac{\partial}{\partial z_1}(\omega \wedge \psi_i \varphi) = d\, v \ , \text{ avec } \ v = u\, d\bar{z}_1 \wedge d\, z_2 \wedge d\,\bar{z}_2 \wedge \dots\, d\, z_n \wedge d\,\bar{z}_n \ ;$$

$$\pi_i^* \, v = \sum_{j=1}^{n} \frac{\beta'_j}{(z'^i_1)^{p_1} \dots (z'^i_n)^{p_n}} \dots \wedge \widehat{dz'^i_j \wedge d\bar{z}'^i_j} \wedge \dots \wedge d\bar{z}'_n$$ (à cause du

type) . Omettons l'indice i pour simplifier la notation.

$$\int_{U'_{\varepsilon'}} \pi^* \frac{\partial}{\partial z_1}(\omega \wedge \psi\varphi) = \int_{U'_{\varepsilon'}} \pi^* dv = \int_{U'_{\varepsilon'}} d\, \pi^* v = \int_{b\, U'_{\varepsilon'}} \pi^* v \ \dots$$

$$= -\sum_{k=1}^{n} \int_{|z'_1| \geq \varepsilon'_1 \dots |z'_k| = \varepsilon'_k \dots |z'_n| \geq \varepsilon'_n} \sum_{j=1}^{n} \frac{\beta'_j}{(z'_1)^{p_1} \dots (z'_n)^{p_n}} \dots \wedge \widehat{dz_j \wedge d\bar{z}_j} \wedge \dots$$

$$\dots \wedge d\bar{z}'_n$$

$$= -\sum_{k=1}^{n} \int_{|z'_1| \geq \varepsilon'_1 \dots |z'_k| = \varepsilon'_n \dots |z'_n| \geq \varepsilon'_n} \frac{\beta'_k}{(z'_1)^{p_1} \dots (z'_n)^{p_n}} \dots \widehat{dz'_k} \wedge d\bar{z}_k \wedge \dots \wedge d\bar{z}'_n ,$$

à cause de la dimension de $|z'_k| = \varepsilon'_k$ dans le plan complexe z'_k .

L'intégrale peut être calculée en intégrant d'abord par rapport à $d\bar{z}'_k$:

$$\int_{|z'_k| = \varepsilon'_k} \beta'_k \frac{d\bar{z}'_k}{(z'_k)^{p_k}} \ ;$$

la limite de cette dernière, quand $\varepsilon'_k \to 0$, est nulle ; il en est de même de : $\lim\limits_{\varepsilon' \to 0} \int_{U'_{\varepsilon'}} \pi^* \frac{\partial}{\partial z_1} (\omega \wedge \psi \varphi)$.

Considérons maintenant :

$$(9) \qquad T'(\pi^*\frac{\partial\omega}{\partial z_1})[\pi^*\varphi] = -\sum_{i=1}^{L} \lim_{\varepsilon'\to o} \int_{U'_{i\varepsilon'}} \pi_i^*\,\omega\wedge\pi_i^*\frac{\partial}{\partial z_1}(\psi_i\varphi) =$$

$$= -\sum_{i=1}^{L} \lim_{\varepsilon'\to o} \int_{\bigcup_{j=1}^{L} U'_{j\varepsilon'}} \pi_i^*\,\omega\wedge\pi_i^*\frac{\partial}{\partial z_1}(\psi_i\varphi) \quad (\text{car } \operatorname{supp}\pi_i^*\frac{\partial}{\partial z_1}(\psi_i\varphi)\subset U'_i)$$

$$= -\lim_{\varepsilon'\to o} \int_{\bigcup_{j=1}^{L} U'_j\,\varepsilon'} \pi^*\omega\wedge\sum_{i=1}^{L}\pi_i^*\frac{\partial}{\partial z_1}(\psi_i\varphi)$$

Mais $\sum_{i=1}^{L}\pi_i^*\frac{\partial}{\partial z_1}(\psi_i\varphi) = \sum_{i=1}^{L}\pi_i^*\psi_i\,\pi_i^*(\frac{\partial}{\partial z_1}\varphi) + \sum_{i=1}^{L}\pi_i^*(\frac{\partial}{\partial z_1}\psi_i)\pi_i^*\varphi$;

cela est égal à

$$(10) \qquad (\sum_{i=1}^{L}\pi_i^*\psi_i)\,\pi^*(\frac{\partial}{\partial z_1}\varphi) + \sum_{i=1}^{L}\pi_i^*(\frac{\partial}{\partial z_1}\psi_i)\pi_i^*\varphi$$

On a : $\pi_i^*\psi_i = \psi'_i$, donc $\sum_{i=1}^{L}\pi_i^*\psi_i = 1$ sur $U\setminus S$; pour tout $j\in[1,\dots,L]$, on a : $\sum_{i=1}^{L}\pi_i^*(\frac{\partial}{\partial z_1}\psi_i)\,\pi_i^*\varphi\,|U'_j\setminus S'_j =$

$= \pi_j^*(\frac{\partial}{\partial z_1}\sum_{i=1}^{L}\psi_i)\,\pi_j^*\varphi|U'_j\setminus S'$; mais $\sum_{i=1}^{L}\psi_i|\pi_j(U'_j\setminus S') = 1$, donc l'expression (10) se réduit à

$$\pi^*\left(\frac{\partial}{\partial z_1}\varphi\right) = \sum_{i=1}^{n} \psi_i' \,\pi^* \frac{\partial}{\partial z_1}\varphi .$$ D'où l'expression de (9) :

$$T'\left(\pi \frac{\partial \omega}{\partial z_1}\right)\left[\pi^* \varphi\right] = -\lim_{\varepsilon' \to o} \int_{L - \bigcup_{j=1}^{L} U'_{j\,\varepsilon'}} \pi^* \omega \wedge \sum_{j=1}^{n} \psi_i' \pi^* \frac{\partial}{\partial z_1}\varphi =$$

$$= -\lim_{\varepsilon' \to o} \sum_{i=1}^{L} \int_{U'_i} \pi^* \omega \wedge \psi_i' \pi^* \frac{\partial}{\partial z_1}\varphi \text{ (car supp } \psi_i' \subset U_i') .$$

$$= -\sum_{i=1}^{L} \psi'_i T'(\pi^*\omega)\left[\pi^* \frac{\partial}{\partial z_1}\varphi\right] = - T'(\pi^*\omega)\left[\pi^* \frac{\partial}{\partial z_1}\varphi\right]$$

$$= - T_\pi(\omega)\left[\frac{\partial}{\partial z_1}\varphi\right] = \frac{\partial T_\pi(\omega)}{\partial z_1}\left[\varphi\right].$$

(b) Pour toute forme $C^\infty \alpha$:

$$\alpha T_\pi(\omega)\left[\varphi\right] = \alpha \pi_* T'(\pi^*\omega) = \pi_* T'(\pi^*\omega)\left[\alpha\varphi\right] = T'(\pi^*\omega)\left[\pi^*(\alpha\varphi)\right] =$$

$$= T'(\pi^*\omega)\left[\pi^*\alpha . \pi^*\varphi\right]$$

$$= T'(\pi^*\alpha\, \pi^*\omega)\left[\pi^*\varphi\right] \text{ (d'après 2.3 (2))}$$

$$= \pi_* T'(\pi^*(\alpha\omega))\left[\varphi\right] = T_\pi(\alpha\omega)\left[\varphi\right].$$

3.7. Lemme : Le courant $T_{\pi*}(\omega)$ est indépendant du morphisme π normalisant pour S .

(a) Si ω est normale, $T_{\pi}(\omega)$, d'après 3.4, 3.5, 3.6 est le courant défini dans le théorème 2.4, donc, il est indépendant de π .

b) Si π est un morphisme : $X' \rightarrow X$ normalisant pour S et $\pi_1 : X'_1 \rightarrow X'$ un morphisme normalisant pour $S' = \pi^{-1}(S)$, soit T' le courant prolongeant ω' canoniquement et soit T'_1 celui qui prolonge canoniquement $\pi_1^* \omega'$; alors, d'après (a) , $\pi_{1*} T'_1 = T'$, donc :

$$T_{\pi}(\omega) = T_{\pi \circ \pi_1}(\omega) .$$

(c) Etant données la variété algébrique projective, irréductible, lisse X sur **C** et la sous-variété algébrique S de X , de codimension 1 , considérons deux variétés algébriques, irréductibles, lisses, birationnellement équivalentes à X et telles qu'il existe dęs morphismes normalisants pour S $\pi_i : X_i \rightarrow X$ (i = 1, 2) ; posons : $S_i = \pi_i^{-1}(S)$. Soit L le corps des fonctions de X , alors le morphisme π_i induit l'isomorphisme θ_i de L sur le corps des fonctions $K(X_i)$ de X_i. Soit $m(L/\mathbf{C})$ la classe des couples (Y, θ) formés d'un **C**-schéma Y et d'un **C**-isomorphisme θ de L sur le corps des fonctions K(Y) de Y .

Dans $m(L/\mathbf{C})$, on dit que (Y', θ') <u>domine</u> (Y'', θ'') si θ' et θ'' étant des isomorphismes de L sur K(Y') et K(Y") respectivement, l'application rationnelle f unique de (Y', θ') dans (Y'', θ'') qui induit l'isomorphisme $\theta' \circ \theta''^{-1} : K(Y'') \rightarrow K(Y')$ est un morphisme. ([3], p. 144) .

Alors, ([3] , p. 144) , il existe $(X'', \theta'') \in m(L/\mathbb{C})$ qui se déduit de (X_1, θ_1) par un nombre fini de transformations monoïdales et tel que (X'', θ'') domine (X_2, θ_2). De plus (X'', θ'') domine (X_1, θ_1) . Enfin , on sait qu'il existe une application rationnelle unique $\pi'' : (X'', \theta'') \to (X, \text{id})$ qui induit l'isomorphisme θ'' ; autrement dit on a le diagramme commutatif suivant :

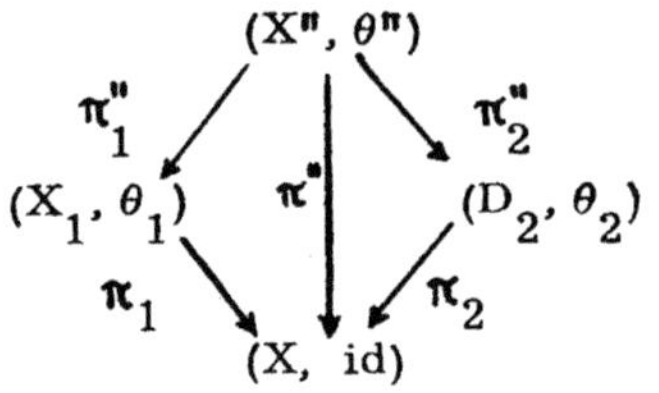

On sait, d'autre part que X_1 étant lisse, il en est de même de X'' . Considérons $\pi''^{-1}(S) = \pi_1''^{-1}(S_1) = \pi''^{-1}(S_2)$. Soit $\pi' : X' \to X''$ un morphisme normalisant pour S'' ; posons $\pi'_i = \pi''_i \circ \pi'$ $(i = 1, 2)$; $\pi_o = \pi''_o \circ \pi'$; alors on le diagramme commutatif de morphismes normalisants pour S, S_1, S_2 resp. :

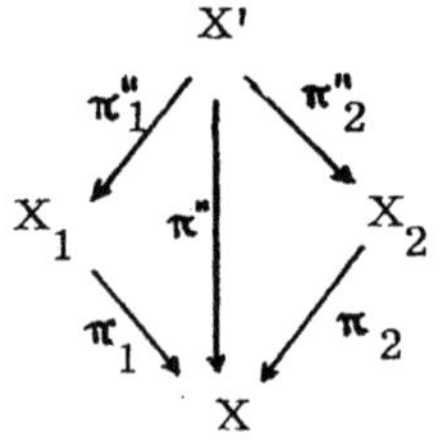

D'après (b), on a : $T_{\pi_1}(\omega) = T_{\pi_1 \circ \pi'}(\omega) = T_{\pi_o}(\omega) = T_{\pi_2 \circ \pi_2}(\omega) = T_{\pi_2}(\omega)$.

Cela établit l'indépendance de $T_{\pi}(\omega)$ par rapport au morphisme π normalisant pour S On désigne désormais le courant $T_{\pi}(\omega)$ par $T(\omega)$.

Résumons les propriétés de $T(\omega)$ dans le théorème suivant :

3.8. Théorème : Soit X une variété algébrique projective, irréductible, lisse sur $\mathbf{C}$ et soit ω une p-forme différentielle semi-méromorphe sur X dont l'ensemble polaire est contenu dans une sous-variété algébrique S de X de codimension 1 . Alors il existe un courant $T(\omega)$ défini canoniquement qui prolonge le courant $\underline{\omega}$ défini par ω sur $X \setminus S$ et qui possède les propriétés suivantes

1) Si ω est à coefficients localement sommables, alors ; $T(\omega)$ est égal au courant défini par ω sur X.

2) L'opérateur $T : \omega \mapsto T(\omega)$ est local et supp $T(\omega) = \operatorname{supp} \omega$.

3) Pour tout opérateur différentiel semi-holomorphe D, on a : $T(D\omega) = DT(\omega)$.

4. Exemple de définition du résidu d'une forme différentielle méromorphe fermée.

On considère le cas d'une forme ω à pôles simples, de degré n sur une variété de dimension complexe n.

4.1. Cas d'une forme élémentaire. On considère une forme ω définie sur le domaine U d'une carte, au voisinage d'un point x de X, de la façon suivante :

$$\omega = \frac{\alpha}{z_{i_1} \dots z_{i_p}} dz_1 \wedge \dots \wedge dz_n$$

où α est holomorphe, alors ω est à coefficients localement sommables et $d\omega = d'\omega = 0$.

Sur U, on considère le courant $T = T(\omega)$; il est défini par:

$$T[\varphi] = \frac{\alpha \wedge \varphi}{z_{i_1} \dots z_{i_p}}$$

$$d''T[\psi] = (-1)^{n+1} T[d''\psi] = \lim_{\varepsilon_1 \to o \dots \varepsilon_n \to o} \int_{|z_1| \geqslant \varepsilon_1 \dots |z_n| \geqslant \varepsilon_n}$$

$$\frac{\alpha dz_1 \wedge \dots \wedge dz_n \wedge d''\psi}{z_{i_1} \dots z_{i_p}} \qquad \psi = \sum_{j=1}^{n} \psi_j \dots \wedge \widehat{d\bar{z}_j} \wedge \dots$$

$$d''\, T[\psi] = -\lim_{\varepsilon_1 \to o \dots \varepsilon_n \to o} \int_{|z_1| \geqslant \varepsilon_1 \dots |z_n| \geqslant \varepsilon_n} d\left(\frac{\alpha dz_1 \wedge \dots \wedge dz_n \wedge \psi}{z_{i_1} \dots z_{i_p}}\right)$$

$$= \lim_{\varepsilon_1 \to o \dots \varepsilon_n \to o} \sum_{k=1}^{n} \int_{|z_1| \geqslant \varepsilon_1 \dots |z_k| = \varepsilon_k \dots} \frac{\alpha \wedge \psi_k}{z_{i_1} \dots z_{i_p}} dz_1 \wedge \dots \wedge dz_n \wedge \dots \wedge \widehat{d\bar{z}_k} \wedge$$

$$= 2\pi i \sum_{l=1}^{p} \int_{S_{i_l}} (\alpha \psi_{i_l} | S_{i_l}) \frac{\ldots \wedge \widehat{dz_{i_l}} \wedge \ldots \wedge \widehat{d\bar{z}_{i_l}} \wedge \ldots}{z_{i_1} \ldots \widehat{z_{i_l}} \ldots z_{i_p}}$$

où S_{i_l} désigne la sous- variété $z_{i_l} = 0$ de U .

$$\frac{\alpha | S_{i_l}}{z_{i_1} \ldots \widehat{z_{i_l}} \ldots z_{i_p}} dz_1 \wedge \ldots \wedge \widehat{dz_{i_l}} \wedge \ldots \wedge dz_n = \text{Rés}_{i_l}(\omega)$$ est appelée la forme résidu de ω sur S_{i_l}

et $$d''T[\psi] = 2\pi i \sum_{l=1}^{p} \int_{S_{i_l}} \text{Rés}_{i_l}(\omega) \wedge (\psi | S_{i_l}) .$$

4.2. **Supposons** que X soit une variété algébrique projective lisse et que ω ait un ensemble polaire contenu dans une sous-variété algébrique S de codimension 1 de X . Soit π un morphisme normalisant pour S ; alors $\pi^* \omega = \omega'$ a ses pôles simples et contenus dans $S' = \pi^{-1}(S)$; $d''T'(\pi^*\omega)$ est défini localement comme en 4.1. et :

$$\pi_* d''T'(\pi^*\omega)[\psi] = d''T'(\pi^*\omega)[\pi^*\psi] = T'(\pi^*\omega)\ [d''\pi^*\psi] = T'(\pi^*\omega)\ [\pi^* d''\psi]$$

$$= \pi_* T'(\pi^*\omega)\ [d''\psi]\ d''T_\pi(\omega)\ [\psi] .$$

$T(\omega)$ étant indépendant de π , on le désignerà par $T(\omega)$ et on appelera $d''T(\omega) = \pi_* d''T'(\pi^*\omega)$ le courant résidu de ω ; il est porté par S .

4.3. Exemple de propriété du courant résidu.

On a: $d'T(\omega) = T(d'\omega) = T(d\omega) = 0$, donc $d''T(\omega) = dT(\omega)$.

Réciproquement, supposons donné un courant t' sur X' , de support contenu dans S' défini localement par

$$(11) \qquad \frac{\alpha \rfloor S'_{i_1} \; dz'_{i_1} \wedge \ldots \wedge \widehat{dz'_{i_1}} \wedge \ldots \wedge dz'_{i_p}}{z'_{i_1} \ldots \check{z}'_{i_1} \ldots z'_{i_p}} \qquad \text{sur } S'_{i_1}$$

(comme $d''T$ au n. 4.1) et soit $t = \pi_* t'$.

Alors , s'il existe un courant L de type $(n, 0)$ sur X tel que: $t = dL$, le courant t est le courant résidu d'une n-forme méromorphe ω sur X .

En effet, pour tout $x \in S$, il existe un voisinage U de x dans X tel que :

$S \cap U = \{\prod_1 \rho_1 = 0 \; ; \rho_1$ fonctions holomorphes sur U irréductibles et distinctes en $x\}$.

Alors, sur $X \setminus S$, on a : $dL = 0$, d'après la définition de t, en particulier $d''L = 0$, donc L est holomorphe sur $X \setminus S$.

Dans le notations de 3.7., désignons par t'_i le courant défini par des expressions de la forme (11) sur U'_i . Alors, au-dessus de U , considérons $\prod_1 \rho_1 L$; on a: $d''(\prod_1 \rho_1 \; L) = \prod_1 \rho_1 \; t = \prod_1 \rho_1 \pi_* t'$.

Pour $\varphi \in \mathcal{D}(U)$, considérons :

$$\prod_1 \rho_1 \pi_* t'[\varphi] = \pi_* t'\left[\prod_1 \rho_1 \varphi\right] = t'\left[\pi^*(\prod_1 \rho_1)\pi^*\varphi\right] .$$

Mais $\pi^*(\prod \rho\ ) = \sum_{i=1}^{L} \psi'_i\, g'_i \prod_k{}' z'^i_k$, où g'_i est une fonction holomorphe sans zéro sur U'_i et $\prod_k{}' z'^i_k$ désigne un produit partiel des coordonnées z'^i_k sur U'_i tel que $S' \cap U'_i = \{\prod_k{}' z'^i_k = 0\}$.

Alors $\prod_1 \rho_1 \pi_* t'[\varphi] = \sum_{i=1}^{L} \psi'_i\, g'_i \prod_k{}' z'^i_k\ t'[\pi^*\varphi] = 0$; donc

$\prod_1 \rho_1$ L est holomorphe au voisinage de x . et L est le courant défini par une n-forme méromorphe ω sur X ayant S comme ensemble polaire à la multiplicité 1 . De plus, $t = d\underline{\omega} = d''\underline{\omega}$, c'est le courant résidu de ω .

P. Dolbeault

BIBLIOGRAPHIE

[1] P.DOLBEAULT, Formes différentielles et cohomologie sur une variété analytique complexe II, Ann. of Math., 65 (1957), 282-330.

[2] P.DOLBEAULT et G. ROBIN, Sur le faisceau des diviseurs à coefficients complexes, C.R. Acad. Sci. Paris, 262 (1966), Série A, 1452-1455.

[2bis] P.DOLBEAULT, Résolution d'un faisceau de formes différentielles méromorphes fermées, Séminaire Lelong, 7e année (1966-67), n.8, 14. p. Paris.

[3] H. HIRONAKA, Resolution of singularities of an algebraic variety over a field of characteristic zero, Ann. of Math., vol. 79 (1964), 109-326.

[4] W.V.D.HODGE and M. ATIYAH, Integrals of the second kind on an algebraic variety, Ann. of Math. 62 (1955), 56-91.

[5] K.KODAIRA, The theorem of Riemann-Roch on compact analytic surfaces, Amer. J. Math., 73 (1951), 813-875.

[6] K.KODAIRA, The theorem of Riemann-Roch for adjoint systems on 3-dimensional algebraic varieties, Ann. of Math. 56 (1952), 298-342.

[7] J.LERAY, Le calcul différentiel et intégral sur une variété analytique complexe (Problème de Cauchy III), Bull. Soc. Math. France, 87 (1959), 81-180.

[8] H. POINCARÉ, Sur les résidus des intégrales doubles, Acta Math. 9 (1887), 321-380.

[9] L.SCHWARTZ, Courant associé à une forme différentielle méromorphe sur une variété analytique complexe, Colloques internationaux du C.N.R.S. : Géométrie différentielle 52, 1953. Strasbourg, 185-195.

CENTRO INTERNAZIONALE MATEMATICO ESTIVO

(C.I.M.E.)

David MUMFORD

(with an appendix by George Kempf)

VARIETIES DEFINED BY QUADRATIC EQUATIONS

Corso tenuto a Varenna dal 9 al 17 settembre 1969

VARIETIES DEFINED BY QUADRATIC EQUATIONS

by David Mumford (with an appendix by George Kempf)
(University of Harvard)

Introduction

First of all, let me fix my terminology and set-up. I will always be working over an algebraically closed ground field k. We will be concerned almost entirely with projective varieties over k (although many of our results generalize immediately to arbitrary projective schemes). By a projective variety, I will understand a topological space X all of whose points are closed, plus a sheaf $\mathcal{O}_X$ of k-valued functions on X isomorphic to some subvariety of $\mathbb{P}^n$ for some n. By a subvariety of $\mathbb{P}^n$, I will mean the subset $X \subset \mathbb{P}^n(k)$ defined by some homogeneous prime ideal $\wp \subset k[X_o,\cdots,X_n]$, with its Zariski-topology and with the sheaf $\mathcal{O}_X$ of functions from X to k induced locally by polynomials in the affine coordinates. Note that our varieties have only k-rational points — no generic points. In this, we depart slightly from the language of schemes. Note too that a projective variety can be isomorphic to many different subvarieties of $\mathbb{P}^n$. An isomorphism of X with a subvariety of $\mathbb{P}^n$ will be called an immersion of X in $\mathbb{P}^n$.

Let me begin with an elementary but somewhat startling result:

Definition: For all d, the d-ple immersion of $\mathbb{P}^n$ is the morphism:

$$s_d: \mathbb{P}^n \longrightarrow \mathbb{P}^N, \qquad N = \binom{n+d}{d} - 1$$

given by:

$$s_d(a_o,\cdots,a_n) = (a^{\alpha^{(o)}},\cdots,a^{\alpha^{(N)}})$$

where $\alpha^{(o)},\ldots,\alpha^{(N)}$ runs through the (n+1)-tuples $\alpha = (\alpha_o,\cdots,\alpha_n)$, such that $\alpha_i \geq 0$, $\Sigma\alpha_i = d$, and

$$a^\alpha = \prod_{i=0}^{n} a_i^{\alpha_i} .$$

Theorem 1: Let $X \subset \mathbb{P}^n$ be a subvariety, and let d_o be the degree of X. For all $d \geq d_o$, consider the new projective embedding:

$$X \subset \mathbb{P}^n \xrightarrow{\ s_d\ } \mathbb{P}^N .$$

Then the subvariety of $\mathbb{P}^N$ so obtained is an intersection of quadrics.*

Proof: Let r = dim X. For all linear spaces L of dimension n-r-2, disjoint from X, let H_L be the join of X and L, i.e., the locus of lines joining X and L. H_L is a hypersurface of degree $\leq d_o$. Then it is easy to see that

$$X = \bigcap_{L\cap X=\emptyset} H_L .$$

In fact, if $x \in \mathbb{P}^n$-X, let

$$\pi:\ \mathbb{P}^n - \{x\} \longrightarrow \mathbb{P}^{n-1}$$

---

*When we talk about an r-dimensional subvariety X of $\mathbb{P}^n$ being an intersection of quadrics, we never mean an intersection of only n-r quadrics (called usually a "complete intersection"). We just mean that there is a large set of quadrics Q_α, $\alpha \in S$, such that $X = \bigcap Q_\alpha$. Of course, S can be assumed finite.

be projection with center x. Then $\pi(X)$ is an r-dimensional subvariety of $\mathbb{P}^{n-1}$ so there exists a linear subspace $M \subset \mathbb{P}^{n-1}$ disjoint from $\pi(X)$ of dimension $(n-1)-r-1$. Choose L such that $\pi(L) = M$. Then $x \notin H_L$.

Thus X is an intersection of hypersurfaces of degree $\leq d_o$. Therefore, for all $d \geq d_o$, X is the intersection of those hypersurfaces of degree d that contain it. But by definition of s_d, if $H_1 \subset \mathbb{P}^n$ is a hypersurface of degree d, there is a hyperplane $H_2 \subset \mathbb{P}^N$ such that

$$H_1 = s_d^{-1}(H_2).$$

Therefore, there is a linear space $K \subset \mathbb{P}^N$ such that $X = s_d^{-1}(K)$, or

$$s_d(X) = K \cap s_d(\mathbb{P}^n).$$

To prove the theorem, it remains to check that $s_d(\mathbb{P}^n)$ is an intersection of quadrics. This follows from the remark:

For all $b_o, \cdots, b_N$,

(*)
$$\left[\begin{array}{l}\text{There exists } a_o, \cdots, a_n \\ \text{such that } b_i = a^{\alpha(i)}\end{array}\right] \Longleftrightarrow \left[\begin{array}{l} b_i b_j = b_k b_\ell \text{ whenever} \\ \alpha(i) + \alpha(j) = \alpha(k) + \alpha(\ell)\end{array}\right]$$

We leave this to the reader.

QED

I want to make 2 remarks. Suppose by the <u>rank</u> r of quadric we mean the rank of the corresponding symmetric matrix. Then the proof of this theorem shows that X is actually an intersection of quadrics of rank ≤ 4. Suppose we make the definition:

<u>Definition</u>: A subvariety $X \subset \mathbb{P}^n$ is <u>ideal-theoretically</u> an intersection of hypersurfaces $H_1, \cdots, H_m$ if set-theoretically:

$$X = H_1 \cap \cdots \cap H_m$$

and moreover, every $x \in X$ has an affine open neighborhood $U \subset \mathbb{P}^n$ such that the ideal $I(X)$ of $X \cap U \subset U$ is generated by the affine equations $f_1, \cdots, f_n$ of $H_1, \cdots, H_n$.

<u>Lemma</u>: If X is non-singular, then X is ideal-theoretically the intersections of $H_1', \cdots, H_n$ if and only if

1) $X = \bigcap_{i=1}^{n} H_i$

2) for all $x \in X$,

$$T_{x,X} = \bigcap_{i=1}^{n} T_{x,H_i}$$

(the intersection being taken in $T_{x,\mathbb{P}^n}$; here T means Zariski tangent space).

We leave the proof to the reader. Using this, we can then prove a variant of Theorem 1 to the effect that if X is non-singular, then $s_d(X)$ is ideal-theoretically an intersection of quadrics.

§1. The cohomological method.

In setting up the concepts of linear systems and ampleness and in the construction of projective embeddings, we have to make a choice between 3 equivalent formulations — that of divisor classes, of line bundles, or of invertible sheaves. It is well known that on any variety X, the group of (Cartier) divisors mod linear equivalence, the group of line bundles and the group of invertible sheaves are all canonically isomorphic. For our purposes, it is most convenient to use the sheaves:

Definition: An invertible sheaf L on X is a sheaf of $\mathcal{O}_X$-modules, locally isomorphic to $\mathcal{O}_X$ itself.

Two such sheaves L_1, L_2 can be tensored to form a 3rd $L_1 \otimes L_2$; $\mathcal{O}_X$ itself is an invertible sheaf forming a unit for this multiplication; and for any L, $L^{-1} = \mathrm{Hom}(L, \mathcal{O}_X)$ is an inverse since $L \otimes L^{-1} \cong L^{-1} \otimes L \cong \mathcal{O}_X$. The set of all invertible sheaves, mod isomorphisms, thus forms an abelian group, called Pic(X).

$\Gamma(L)$ or $H^0(L)$ will be the vector space of global sections of L. If $s \in \Gamma(L)$, and $x \in X$, then via an isomorphism $L|_U \cong \mathcal{O}_X|_U$ in some neighborhood U of x, we can find a value s(x); and the conditions $s(x) = 0$ or $s(x) \neq 0$ are independent of this local isomorphism.

<u>Definition</u>: The <u>base points</u> of $\Gamma(L)$ are the points $x \in X$ such that for all $s \in \Gamma(L)$, $s(x) = 0$.

If $\Gamma(L)$ is base point free, L defines a canonical morphism into projective space. Let $\mathbb{P}(\Gamma(L))$ be the projective space of hyperplanes in $\Gamma(L)$. Then define

$$\phi_L : X \longrightarrow \mathbb{P}(\Gamma(L))$$

by $$\phi_L(x) = \{s \in \Gamma(L) \mid s(x) = 0\}.$$

This is easily checked to be a morphism. More explicitly, let $s_o, s_1, \cdots, s_n$ be a basis of $\Gamma(L)$. Define:

$$\phi_L : X \longrightarrow \mathbb{P}^n$$

by $$\phi_L(x) = \text{pt. with homog. coord. } (s_o(x), s_1(x), \cdots, s_n$$

<u>Definition</u>: L is <u>very ample</u> if $\Gamma(L)$ is base point free and ϕ_L is an immersion (= an isomorphism of X with $\phi_L(X)$). L is <u>ample</u> if L^n is very ample for some $n \geq 1$.

Write $\mathbb{P}$ for $\mathbb{P}(\Gamma(L))$ and suppose L is very ample. Then the vector space $\Gamma(L)$ is <u>canonically</u> isomorphic to the space of homogeneo coordinate functions on the projective space $\mathbb{P}$, i.e,

$$\Gamma(L) \cong \Gamma(\mathbb{P}, \mathcal{O}_{\mathbb{P}}(1)).$$

And the kth symmetric power of $\Gamma(L)$, which we write $S^k\Gamma(L)$, is canonically isomorphic to the space of homogeneous polynomials of degree k in the homogeneous coordinates on $\mathbb{P}$, i.e.,

$$S^k\Gamma(L) \cong \Gamma(\mathbb{P}, \mathcal{O}_{\mathbb{P}}(k)).$$

Thus the vector space of homogeneous polynomials of degree k that vanish on $\phi_L(X)$ is nothing but the kernel of the canonical map:

$$S^k\Gamma(L) \longrightarrow \Gamma(L^k).$$

A strengthening of the assertion that $\phi_L(X)$ is an intersection of quadrics is that its homogeneous ideal is generated by quadrics. This is the same as asking whether the canonical map:

$$S^{k-2}\Gamma(L) \otimes \mathrm{Ker}\left[S^2\Gamma(L) \to \Gamma(L^2)\right] \longrightarrow \mathrm{Ker}\left[S^k\Gamma(L) \longrightarrow \Gamma(L^k)\right]$$

is surjective for all $k \geq 2$.

Our basic definition is this:

<u>Definition</u>: Let $\mathcal{F}, \mathcal{G}$ be coherent sheaves on X. Define $\mathcal{R}(\mathcal{F},\mathcal{G})$, $\mathcal{S}(\mathcal{F},\mathcal{G})$ as the kernel and cokernel of the canonical map α:

$$0 \longrightarrow \mathcal{R}(\mathcal{F},\mathcal{G}) \longrightarrow \Gamma(\mathcal{F})\otimes\Gamma(\mathcal{G}) \xrightarrow{\alpha} \Gamma(\mathcal{F}\otimes\mathcal{G}) \longrightarrow \mathcal{S}(\mathcal{F},\mathcal{G}) \longrightarrow 0.$$

Thus if L is a very ample invertible sheaf, $\mathcal{R}(L,L)$ is the space (a) of alternating elements of $\Gamma(L)\otimes\Gamma(L)$, and (b) of the quadratic relations holding on $\phi_L(X)$.

<u>Definition</u>: Let L be an ample sheaf on X. Then L is <u>normally generated</u> if

$$\Gamma(L)^{\otimes k} \longrightarrow \Gamma(L^k)$$

is surjective, all $k \geq 1$.

This is clearly equivalent to the condition $\mathcal{S}(L^i, L^j) = (0)$, $i,j \geq 1$. Note that if L is normally generated then L is necessarily very ample too! In fact, consider the 2 morphisms:

$$\begin{array}{ccc} & & \mathbb{P}(\Gamma(L^n)) \\ & \nearrow^{\phi_{L^n}} & \\ X & & \\ & \searrow_{\phi_L} & \\ & & \mathbb{P}(\Gamma(L)) \end{array}$$

The n-ple embedding of the projective space $\mathbb{P}(V)$ of hyperplanes for <u>any</u> vector space V is canonically a map

$$s_n : \mathbb{P}(V) \longrightarrow \mathbb{P}(S^n V).$$

Moreover, via the surjection

$$S^n\Gamma(L) \longrightarrow \Gamma(L^n),$$

we can identify $\mathbb{P}(\Gamma(L^n))$ canonically with a linear subspace of $\mathbb{P}(S^n\Gamma(L))$. Putting this together, we get a diagram:

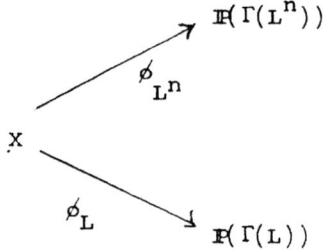

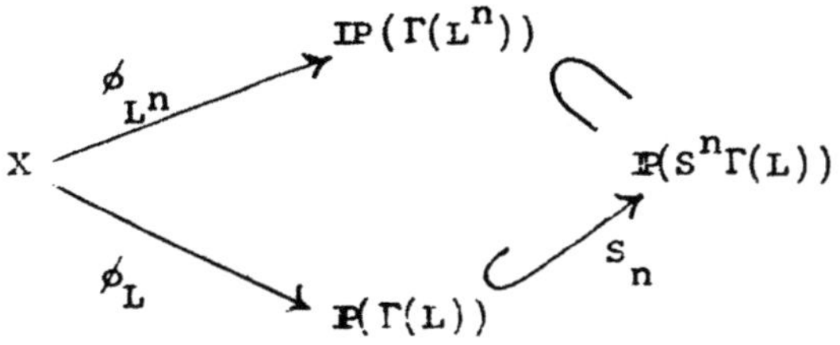

It is easy to check that this commutes. Now for large n, L^n is very ample, hence ϕ_{L^n} is an immersion, so it follows from the diagram that ϕ_L is an immersion too, i.e., L is very ample.

__Definition__: Let L be a normally generated invertible sheaf. Then L is __normally presented__ if one of the 4 equivalent conditions holds:

(A) $\text{Ker}[S^2\Gamma(L) \longrightarrow \Gamma(L^2)] \otimes \Gamma(L^{k-2}) \longrightarrow \text{Ker}[S^k\Gamma(L) \longrightarrow \Gamma(L^k)]$

is surjective, all $k \geq 2$

(B) $\bigoplus_{1 \leq i < j \leq n} [\mathcal{R}(L,L) \otimes \Gamma(L)^{k-2}] \longrightarrow \text{Ker}[\Gamma(L)^{\otimes k} \longrightarrow \Gamma(L^k)]$

is surjective, all $k \geq 2$.

The above homomorphism maps an element $a \otimes b$ in the $(i,j)^{th}$ factor to the element of $\Gamma(L)^{\otimes k}$ whose i^{th} and j^{th} components are determined by a, and the rest by b.

(C) $\Gamma(L^{i-1}) \otimes \mathcal{R}(L,L) \otimes \Gamma(L^{j-1}) \longrightarrow \mathcal{R}(L^i, L^j)$

is surjective, if $i,j \geq 1$.

Here, if $\Sigma a_i \otimes b_i \in \mathcal{R}(L,L) \subset \Gamma(L) \otimes \Gamma(L)$, and $c \in \Gamma(L^{i-1})$, $d \in \Gamma(L^{j-1})$, then we map $c \otimes (\Sigma a_i \otimes b_i) \otimes d$ to $\Sigma(a_i c) \otimes (b_i d) \in \Gamma(L^i) \otimes \Gamma(L^j)$.

(D) $\mathcal{R}(L^i,L^j) \otimes \Gamma(L^k) \longrightarrow \mathcal{R}(L^i,L^{j+k})$

is surjective if $i,j,k \geq 1$.

It is not so obvious that all these properties are equivalent! Thus to see (A) $\Longleftrightarrow$ (B), note that $\mathcal{R}(L,L) \subset \Gamma(L) \otimes \Gamma(L)$ contains the alternating tensors, so the image of

$$\bigoplus_{1 \leq i < j \leq n} [\mathcal{R}(L,L) \otimes \Gamma(L)^{k-2}]$$

in $\Gamma(L)^k$ contains all the alternating tensors. So the image equals $\mathrm{Ker}(\Gamma(L)^k \longrightarrow \Gamma(L^k))$ if and only if its image in $S^k\Gamma(L)$ equals $\mathrm{Ker}(S^k\Gamma(L) \longrightarrow \Gamma(L^k))$. But its image in $S^k\Gamma(L)$ is the same as the image of the map in (A).

(C) $\Longrightarrow$ (D) follows immediately using normal generation and (D) $\Longrightarrow$ (C) follows by factoring the map in (C) thus:

$$\Gamma(L^{i-1}) \otimes \mathcal{R}(L,L) \otimes \Gamma(L^{j-1}) \longrightarrow \Gamma(L^{i-1}) \otimes \mathcal{R}(L,L^j) \longrightarrow \mathcal{R}(L^i,L^j)$$

Next, to prove (C) $\Longrightarrow$ (B), factor $\Gamma(L)^k \longrightarrow \Gamma(L^k)$ as follows:

$$\Gamma(L) \otimes \Gamma(L)^{k-1} \xrightarrow{\text{onto}} \Gamma(L^2) \otimes \Gamma(L)^{k-2} \xrightarrow{\text{onto}} \cdots \xrightarrow{\text{onto}} \Gamma(L^k) .$$

To prove (B), it is enough to show that $\oplus[\mathcal{R}(L,L)\otimes\Gamma(L)^{k-2}]$ goes onto the kernel at each stage of this sequence. Thus it is enough if $\Gamma(L)^{i-1} \otimes \mathcal{R}(L,L)$ is mapped onto $\mathrm{Ker}[\Gamma(L^i) \otimes \Gamma(L) \longrightarrow \Gamma(L^{i+1})]$. This last space is $\mathcal{R}(L^i,L)$, so this ontoness is part of (C).

Finally, to prove (B) $\Longrightarrow$ (C), factor $\Gamma(L)^k \longrightarrow \Gamma(L^k)$ when $k = i+j$, as follows:

$$\Gamma(L)^{i+j} \xrightarrow[\alpha]{\text{onto}} \Gamma(L^i) \otimes \Gamma(L^j) \xrightarrow[\beta]{\text{onto}} \Gamma(L^{i+j}).$$

It follows from normal generation that we get a surjection:

$$\mathrm{Ker}(\beta\cdot\alpha) \xrightarrow{\text{onto}} \mathrm{Ker}(\beta) = \mathcal{R}(L^i,L^j).$$

But $\mathrm{Ker}(\beta\,\alpha)$ is generated by $\oplus[\mathcal{R}(L,L) \otimes \Gamma(L)^{i+j-2}]$. The image of this last space in $\Gamma(L^i)\otimes\Gamma(L^j)$ is the same as the image of $\Gamma(L^{i-1}) \otimes\mathcal{R}(L,L) \otimes \Gamma(L^{j-1})$, so (C) follows.

This at least gives us a nice definition to work with! It seems easier to prove things about $\mathcal{S}$ first, and then to use these results to obtain things about $\mathcal{R}$. Our first result is:

<u>Theorem 2</u> (Generalized lemma of Castelnuovo): Suppose L is an ample invertible sheaf on a variety X such that $\Gamma(L)$ has no base points. Suppose $\mathfrak{F}$ is a coherent sheaf on X such that

$$H^i(\mathfrak{F} \otimes L^{-i}) = (0), \quad i \geq 1.$$

Then (a) $H^i(\mathfrak{F} \otimes L^j) = (0)$ if $i+j \geq 0$, $i \geq 1$

and (b) $\mathcal{S}(\mathfrak{F} \otimes L^i, L) = (0)$, $i \geq 0$.

To motivate this, look at the case of Castelnuovo's original lemma: X = non-singular curve, $\mathfrak{A}$, $\mathfrak{B}$ divisors on X, $\mathfrak{F} = \mathcal{O}_X(\mathfrak{B})$, $L = \mathcal{O}_X(\mathfrak{A})$. In classical language:

$$\begin{pmatrix}\Gamma(L) \text{ has no} \\ \text{base points}\end{pmatrix} \underset{\text{def}}{\equiv} \begin{pmatrix}|\mathfrak{A}| \text{ is base point} \\ \text{free}\end{pmatrix}$$

$$\left(H^1(\mathfrak{F} \otimes L^{-1}) = (0)\right) \underset{\text{def}}{\equiv} \left(|\mathfrak{B} - \mathfrak{A}| \text{ is non-special}\right)$$

Translating the conclusion, we find:

$$\left(\mathcal{S}(\mathfrak{F},L) = (0)\right) \underset{\text{def}}{\equiv} \left(|\mathfrak{A}+\mathfrak{B}| = \begin{matrix}\text{the minimal sum} \\ |\mathfrak{A}| + |\mathfrak{B}|\end{matrix}\right)$$

<u>Proof of Theorem 2</u>: Use induction on dim(Supp $\mathfrak{F}$). If dim(Supp $\mathfrak{F}$) = 0, then choose $s \in \Gamma(L)$ such that $s(x) \neq 0$ for all $x \in \text{Supp}(\mathfrak{F})$. Then

$$\Gamma(\mathfrak{F}) \underset{k}{\otimes} (s^i \cdot k) \xrightarrow{\approx} \Gamma(\mathfrak{F} \otimes L^i)$$

is an isomorphism, so certainly

$$\Gamma(\mathfrak{F}) \underset{k}{\otimes} \Gamma(L^i) \longrightarrow \Gamma(\mathfrak{F} \otimes L^i)$$

is surjective. Therefore $\mathcal{S}(\mathfrak{F},L^i) = (0)$. Also, all groups $H^i(\mathfrak{F} \otimes \text{anything})$, $i \geq 1$, vanish.

Now suppose we are given an $\mathfrak{F}$, and we have proven the theorem for all $\mathfrak{F}^*$'s with $\dim(\text{Supp }\mathfrak{F}^*) < \dim(\text{Supp }\mathfrak{F})$. I claim that there is an element $s \in \Gamma(L)$ sufficiently "generic" so that for every $x \in X$, if we choose an isomorphism $L|_U \cong \mathcal{O}_X|_U$ near x, so that s can be considered as a function, then s is <u>not</u> a 0-divisor in the stalk $\mathfrak{F}_x$ of $\mathfrak{F}$. To see, recall that by the Noetherian decomposition theorems, for any coherent $\mathfrak{F}$, there is a finite set of irreducible subsets $Z_1, \cdots, Z_n \subset \mathrm{Supp}(\mathfrak{F})$ (including the components of $\mathrm{Supp}(\mathfrak{F})$, but possibly including some "embedded components" too) such that the support of any element

$$\alpha \in \Gamma(U, \mathfrak{F})$$

is a union of some of the sets $U \cap Z_i$. For each i, not all sections $s \in \Gamma(L)$ vanish identically on Z_i. Therefore there is an element $s \in \Gamma(L)$ not identically zero on any Z_i. If $\alpha \in \Gamma(U, \mathfrak{F})$, then s must be non-zero at at least one point x of $\mathrm{Supp}(\alpha)$, hence $\alpha \otimes s \in \Gamma(U, \mathfrak{F} \otimes L)$ is not zero near x. Thus s has the required property and the map $\mathfrak{F} \longrightarrow \mathfrak{F} \otimes L$, defined by $\alpha \longmapsto \alpha \otimes s$, is injective.

It is more convenient to use the map $\mathfrak{F} \otimes L^{-1} \longrightarrow \mathfrak{F}$, defined by $\alpha \longmapsto \alpha \otimes s$. Let $\mathfrak{F}^*$ be the cokernel. Then for all i, we have exact sequences:

$$(*)_i \qquad 0 \longrightarrow \mathfrak{F} \otimes L^{-i-1} \xrightarrow{\ \otimes s\ } \mathfrak{F} \otimes L^{-i} \longrightarrow \mathfrak{F}^* \otimes L^{-i} \longrightarrow 0.$$

Note that $\dim(\mathrm{Supp}\,\mathfrak{F}^*) < \dim(\mathrm{Supp}\,\mathfrak{F})$. In fact, for all i, $\otimes s$ is an isomorphism on almost all of Z_i, hence $Z_i \not\subset \mathrm{Supp}(\mathfrak{F}^*)$. Therefore every component of $\mathrm{Supp}(\mathcal{F}^*)$ is a proper closed subset of some component of $\mathrm{Supp}(\mathcal{F})$. By $(*)_i$, we get an exact sequence:

$$\begin{array}{ccccc} H^i(\mathcal{F}\otimes L^{-i}) & \longrightarrow & H^i(\mathcal{F}^*\otimes L^{-i}) & \longrightarrow & H^{i+1}(\mathcal{F}\otimes L^{-i-1}), \qquad i \geq 1 \\ \| & & & & \| \\ (0) & & & & (0) \end{array}$$

hence $H^i(\mathcal{F}^*\otimes L^{-i}) = (0)$. Thus the hypothesis of the theorem is valid for $\mathcal{F}^*$, so by our induction hypothesis, so is the conclusion. Going back from $\mathcal{F}^*$ to $\mathcal{F}$, use the exact sequence:

$$H^i(\mathcal{F} \otimes L^{-i}) \longrightarrow H^i(\mathcal{F} \otimes L^{-i+1}) \longrightarrow H^i(\mathcal{F}^* \otimes L^{-i+1}).$$

The 1st group is (0) by the hypothesis on $\mathcal{F}$; the 3rd group is (0) by the theorem for $\mathcal{F}^*$: so the 2nd is (0). Replacing $\mathcal{F}$ by $\mathcal{F} \otimes L$, we continue in this way and prove by induction on $i+j$ that

$$H^i(\mathcal{F} \otimes L^j) = (0), \quad i+j \geq 0, \quad i \geq 1.$$

As for the $\mathcal{S}$'s, look at the diagram of solid arrows:

$$\begin{array}{ccccccccc}
0 \longrightarrow & \Gamma(\mathcal{F}\otimes L^{-1}) \otimes \Gamma(L) & \longrightarrow & \Gamma(\mathcal{F}) \otimes \Gamma(L) & \longrightarrow & \Gamma(\mathcal{F}^*) \otimes \Gamma(L) & \longrightarrow 0 \\
& \downarrow & & \downarrow & & \downarrow & \\
0 \longrightarrow & \Gamma(\mathcal{F}) & \longrightarrow & \Gamma(\mathcal{F}\otimes L) & \longrightarrow & \Gamma(\mathcal{F}^*\otimes L) & \longrightarrow 0 \\
& \downarrow & & \downarrow & & \downarrow & \\
& \mathcal{S}(\mathcal{F}\otimes L^{-1}, L) & \xrightarrow{\alpha} & \mathcal{S}(\mathcal{F}, L) & \longrightarrow & \mathcal{S}(\mathcal{F}^*, L) & \longrightarrow 0 \\
& \downarrow & & \downarrow & & \downarrow & \\
& 0 & & 0 & & 0 &
\end{array}$$

It has exact rows since $H^1(\mathfrak{I}\otimes L^{-1}) = H^1(\mathfrak{I}) = (0)$, and exact columns by definition. Define the dotted arrow by $\alpha \longmapsto \alpha\otimes s$. Then the shaded triangle commutes, which proves that the map α is zero! Since $\mathcal{S}(\mathfrak{I}^*,L) = (0)$, it follows that $\mathcal{S}(\mathfrak{I},L) = (0)$. As we may replace $\mathfrak{I}$ by $\mathfrak{I}\otimes L^i$, $i \geq 1$, the rest of (b) follows too.

QED

A useful remark is that a close examination of this proof shows a slightly more precise result. Namely, that if $n = \dim(\mathrm{Supp}\ \mathfrak{I})$; and if $s_o,\cdots,s_n \in \Gamma(L)$ are sufficiently "generic" elements, then in fact $\Gamma(\mathfrak{I}\otimes L)$ is spanned by the images of $\Gamma(\mathfrak{I}) \otimes_k (s_i \cdot k)$, for $0 \leq i \leq n$.

<u>Theorem 3</u>: Let L be an ample invertible sheaf on an n-dimensional variety X. Suppose $\Gamma(L)$ has no base points and

$$H^i(L^j) = (0), \quad i \geq 1,\ j \geq 1.$$

Then $\mathcal{S}(L^i,L^j) = (0)$ if $i \geq n+1$, $j \geq 1$.

In particular, if $i \geq n+1$, L^i is ample with normal generation, hence very ample.

<u>Proof</u>: Apply Theorem 2 to $\mathfrak{I} = L^{n+1}$. It follows that $\mathcal{S}(L^i,L) = (0)$, if $i \geq n+1$. Explicitly $\Gamma(L^i) \otimes \Gamma(L) \longrightarrow \Gamma(L^{i+1})$ is surjective if $i \geq n+1$. Composing these maps, $\Gamma(L^i) \otimes \Gamma(L)^j \longrightarrow \Gamma(L^{i+j})$ is surjective if $i \geq n+1$. Therefore $\Gamma(L^i) \otimes \Gamma(L^j) \longrightarrow \Gamma(L^{i+j})$ is surjective too, if $i \geq n+1$.

QED

Next we want to prove similar results about $\mathcal{R}$. We need the preliminary result:

<u>6-lemma</u>: If $0 \longrightarrow \mathfrak{F}_1 \longrightarrow \mathfrak{F}_2 \longrightarrow \mathfrak{F}_3 \longrightarrow 0$ is an exact sequence of coherent sheaves, and $\Gamma(\mathfrak{F}_2) \longrightarrow \Gamma(\mathfrak{F}_3)$ is surjective — e.g., if $H^1(\mathfrak{F}_1) = (0)$ — then for all invertible sheaves L there is an exact sequence:

$$0 \longrightarrow \mathcal{R}(\mathfrak{F}_1, L) \longrightarrow \mathcal{R}(\mathfrak{F}_2, L) \longrightarrow \mathcal{R}(\mathfrak{F}_3, L) \longrightarrow \mathcal{S}(\mathfrak{F}_1, L) \longrightarrow \mathcal{S}(\mathfrak{F}_2, L) \longrightarrow \mathcal{S}(\mathfrak{F}_3, L$$

Also, even if $\Gamma(\mathfrak{F}_2) \longrightarrow \Gamma(\mathfrak{F}_3)$ is not surjective, the 1st 3 terms form an exact sequence.

<u>Proof</u>: Look at the diagram of solid arrows:

$$\begin{array}{ccccccccc}
 & & 0 & & 0 & & 0 & & \\
 & & \downarrow & & \downarrow & & \downarrow & & \\
0 & \longrightarrow & \mathcal{R}(\mathfrak{F}_1, L) & \dashrightarrow & \mathcal{R}(\mathfrak{F}_2, L) & \dashrightarrow & \mathcal{R}(\mathfrak{F}_3, L) & & \\
 & & \downarrow & & \downarrow & & \downarrow & & \\
0 & \longrightarrow & \Gamma(\mathfrak{F}_1)\otimes\Gamma(L) & \longrightarrow & \Gamma(\mathfrak{F}_2)\otimes\Gamma(L) & \longrightarrow & \Gamma(\mathfrak{F}_3)\otimes\Gamma(L) & \longrightarrow & 0 \\
 & & \downarrow & & \downarrow & & \downarrow & & \\
0 & \longrightarrow & \Gamma(\mathfrak{F}_1\otimes L) & \longrightarrow & \Gamma(\mathfrak{F}_2\otimes L) & \longrightarrow & \Gamma(\mathfrak{F}_3\otimes L) & & \\
 & & \downarrow & & \downarrow & & \downarrow & & \\
 & \dashrightarrow & \mathcal{S}(\mathfrak{F}_1, L) & \dashrightarrow & \mathcal{S}(\mathfrak{F}_2, L) & \dashrightarrow & \mathcal{S}(\mathfrak{F}_3, L) & & \\
 & & \downarrow & & \downarrow & & \downarrow & & \\
 & & 0 & & 0 & & 0 & &
\end{array}$$

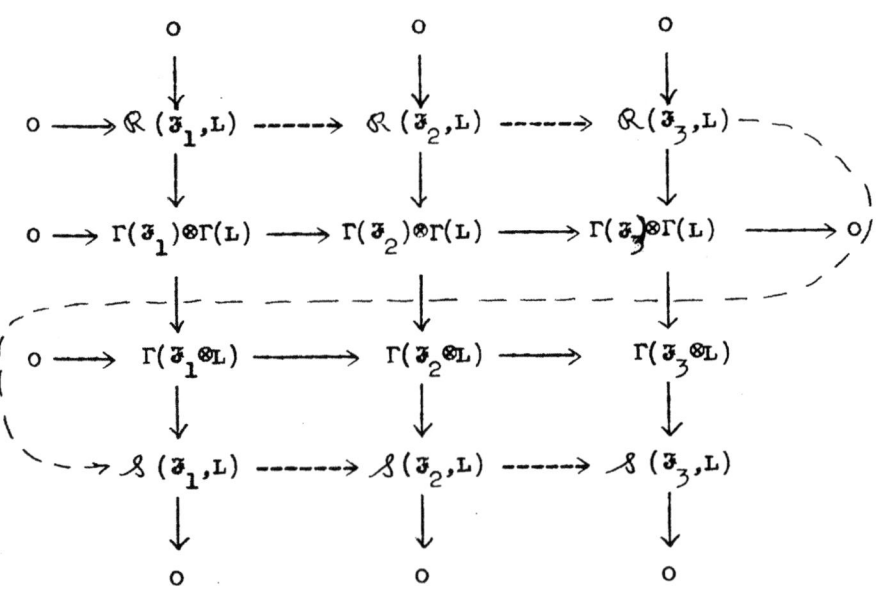

The rows and columns are exact, by hypothesis. By the so-called "serpent" argument, you get an exact sequence indicated by the dotted arrows.

QED

We apply this to prove:

Theorem 4: Let L and M be ample invertible sheaves on a projective variety X, let $\mathfrak{F}$ be a coherent sheaf on X, and assume:

i) $\Gamma(L)$, $\Gamma(M)$ have no base points,

ii) $H^{i+j-1}(\mathfrak{F}\otimes L^{-i}\otimes M^{-j}) = (0)$ if $i,j \geq 1$.

Then the natural map:

$$\mathcal{R}(\mathfrak{F},L) \otimes \Gamma(M) \longrightarrow \mathcal{R}(\mathfrak{F} \otimes M, L)$$

is surjective.

[One can also check that the hypotheses imply that

$$H^k(\mathfrak{F}\otimes L^{-i}\otimes M^{-j}) = (0) \text{ if } k \geq 1,\ i+k \geq 0,\ j+k \geq 0,\ i+j+k \geq -1.$$

Therefore the hypotheses are stable under the substitution $\mathfrak{F} \longmapsto \mathfrak{F} \otimes L$ or $\mathfrak{F} \otimes M$. However, we may as well stick to the simplest case of the theorem.]

Proof: As in Theorem 2, we use induction on $\dim(\mathrm{Supp}\ \mathfrak{F})$. If $\dim(\mathrm{Supp}\ \mathfrak{F}) = 0$, we get the diagram:

$$\begin{array}{ccccc}
0 & & 0 & & 0 \\
\downarrow & & \downarrow & & \downarrow \\
\mathcal{R}(\mathfrak{J},L)\otimes\Gamma(M) & \longrightarrow & \mathcal{R}(\mathfrak{J}\otimes M, L) & \underset{\alpha}{\cong} & \mathcal{R}(\mathfrak{J},L) \\
\downarrow & & \downarrow & & | \\
\Gamma(\mathfrak{J})\otimes\Gamma(L)\otimes\Gamma(M) & \longrightarrow & \Gamma(\mathfrak{J}\otimes M)\otimes\Gamma(L) & \underset{\beta}{\cong} & \Gamma(\mathfrak{J})\otimes\Gamma(L) \\
\downarrow & & \downarrow & & \downarrow \\
\Gamma(\mathfrak{J}\otimes L)\otimes\Gamma(M) & \longrightarrow & \Gamma(\mathfrak{J}\otimes L\otimes M) & \underset{\gamma}{\cong} & \Gamma(\mathfrak{J}\otimes L)
\end{array}$$

where the isomorphisms β and γ are obtained by choosing a section $s \in \Gamma(M)$ non-zero at all points of $\mathrm{Supp}(\mathfrak{J})$, hence an isomorphism of M and $\mathcal{O}_X$ in a neighborhood of $\mathrm{Supp}(\mathfrak{J})$. β and γ induce an isomorphism α. But in the map on the top row, if $p \in \mathcal{R}(\mathfrak{J},L)$, then $p\otimes s \in \mathcal{R}(\mathfrak{J},L)\otimes\Gamma(M)$ is taken to $p \in \mathcal{R}(\mathfrak{J},L)$, so this map is surjective. This proves the theorem when $\dim(\mathrm{Supp}\,\mathfrak{J}) = 0$.

In the general case, choose a good section $s \in \Gamma(M)$ as in the proof of Theorem 2 so as to obtain an exact sequence:

$$0 \longrightarrow \mathfrak{J}\otimes M^{-1} \xrightarrow{\otimes s} \mathfrak{J} \longrightarrow \mathfrak{J}^* \longrightarrow 0$$

with $\dim(\mathrm{Supp}\,\mathfrak{J}^*) < \dim(\mathrm{Supp}\,\mathfrak{J})$. We obtain exact sequences:

$$H^{i+j-1}(\mathfrak{J}\otimes L^{-i}\otimes M^{-j}) \longrightarrow H^{i+j-1}(\mathfrak{J}^*\otimes L^{-i}\otimes M^{-j}) \longrightarrow H^{i+j}(\mathfrak{J}\otimes L^{-i}\otimes M^{-j-1}).$$

The 1st and 3rd groups are 0 by hypothesis, so the 2nd is also. Thi shows that $\mathfrak{J}^*$ satisfies the hypotheses of the theorem too. So by the induction hypothesis, $\mathcal{R}(\mathfrak{J}^*,L)\otimes\Gamma(M) \longrightarrow \mathcal{R}(\mathfrak{J}^*\otimes M, L)$ is surjective. Moreover, by Castelnuovo's lemma (Theorem 2), applied t

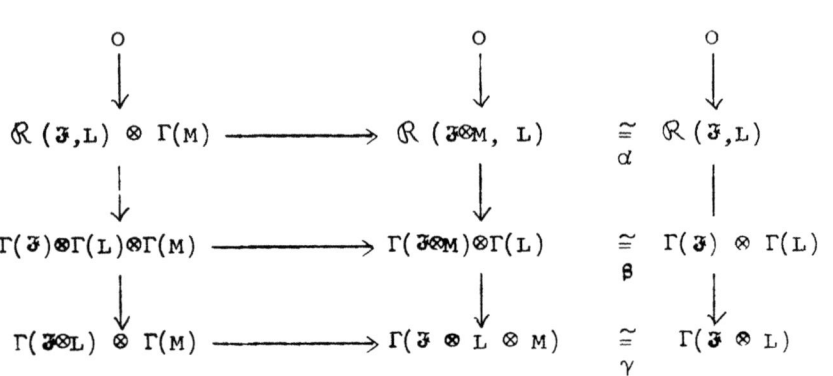

$\mathfrak{F} \otimes M^{-1}$ and L, $\mathcal{S}(\mathfrak{F} \otimes M^{-1}, L) = (0)$ and $H^1(\mathfrak{F} \otimes M^{-1}) = (0)$. Applying the 6-lemma, we deduce that:

$$0 \longrightarrow \mathcal{R}(\mathfrak{F} \otimes M^{-1}, L) \longrightarrow \mathcal{R}(\mathfrak{F}, L) \longrightarrow \mathcal{R}(\mathfrak{F}^*, L) \longrightarrow 0$$

is exact. Now consider the diagram of solid arrows:

$$\begin{array}{ccccccccc}
0 \longrightarrow & \mathcal{R}(\mathfrak{F} \otimes M^{-1}, L) \otimes \Gamma(M) & \longrightarrow & \mathcal{R}(\mathfrak{F}, L) \otimes \Gamma(M) & \longrightarrow & \mathcal{R}(\mathfrak{F}^*, L) \otimes \Gamma(M) & \longrightarrow 0 \\
& \downarrow & & \downarrow \beta & & \gamma \downarrow & \\
0 \longrightarrow & \mathcal{R}(\mathfrak{F}, L) & \xrightarrow[\alpha]{} & \mathcal{R}(\mathfrak{F} \otimes M, L) & \longrightarrow & \mathcal{R}(\mathfrak{F}^* \otimes M, L) &
\end{array}$$

If you define the dotted arrow by $a \longmapsto a \otimes s$, it is clear that the shaded triangle commutes. Therefore $\mathrm{Im}(\alpha) \subset \mathrm{Im}(\beta)$ and using the surjectivity of γ, the surjectivity of β follows.

QED

To apply this Theorem, we need another result:

Proposition: Let $\mathfrak{F}$ be a coherent sheaf, and L, M invertible sheaves on X. If

a) $\mathcal{R}(\mathfrak{F}, L) \otimes \Gamma(M) \longrightarrow \mathcal{R}(\mathfrak{F} \otimes M, L)$ is surjective

b) $\mathcal{S}(\mathfrak{F}, L) = (0)$,

then

c) $\mathcal{R}(\mathfrak{F}, M) \otimes \Gamma(L) \longrightarrow \mathcal{R}(\mathfrak{F} \otimes L, M)$ is surjective.

Proof: Use the diagram:

$$\begin{array}{ccccccc}
 & & & & 0 & & 0 \\
 & & & & \downarrow & & \downarrow \\
 & & & & \Gamma(L) \otimes \mathcal{R}(\mathfrak{J},M) & \xrightarrow{\gamma} & \mathcal{R}(L \otimes \mathfrak{J}, M) \\
 & & & & \downarrow & & \downarrow \\
0 & \longrightarrow & \mathcal{R}(L,\mathfrak{J}) \otimes \Gamma(M) & \longrightarrow & \Gamma(L) \otimes \Gamma(\mathfrak{J}) \otimes \Gamma(M) & \xrightarrow{\beta} & \Gamma(L \otimes \mathfrak{J}) \otimes \Gamma(M) \\
 & & \downarrow{\scriptstyle\alpha} & & \downarrow & & \downarrow \\
0 & \longrightarrow & \mathcal{R}(L, \mathfrak{J} \otimes M) & \longrightarrow & \Gamma(L) \otimes \Gamma(\mathfrak{J} \otimes M) & \longrightarrow & \Gamma(L \otimes \mathfrak{J} \otimes M)
\end{array}$$

By assumption, α and β are surjective. "Chasing" the diagram, one sees quickly that γ is surjective too.

QED

Theorem 5: Let L be an ample invertible sheaf on an n-dimensional variety X. Assume:

i) $\Gamma(L)$ is base point free,

ii) $H^i(L^j) = (0)$ if $i,j \geq 1$.

Then it follows that:

$$\mathcal{R}(L^i, L^j) \otimes \Gamma(L^k) \longrightarrow \mathcal{R}(L^{i+k}, L^j)$$

is surjective, if $i \geq n+2$, $j,k \geq 1$. In particular, if $i \geq n+2$, L^i is ample with normal presentation.

Proof: By Theorem 4,

$$\mathcal{R}(L^i, L) \otimes \Gamma(L) \longrightarrow \mathcal{R}(L^{i+1}, L)$$

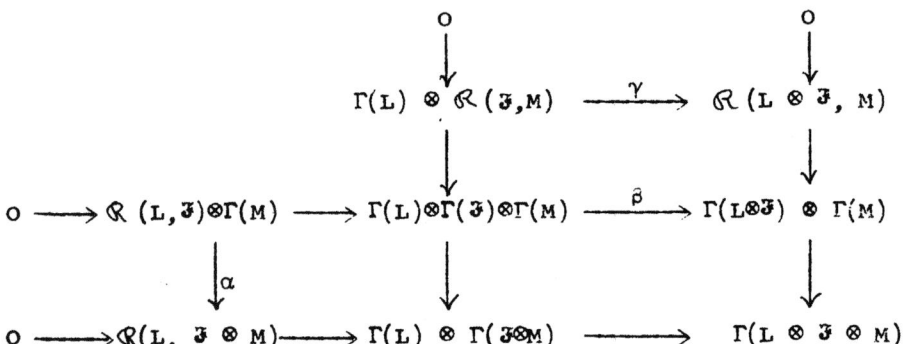

is surjective, if $i \geq n+2$. Iterating, we find that:

$$\mathcal{R}\,(L^i,L) \otimes \Gamma(L^j) \longrightarrow \mathcal{R}\,(L^{i+j},L)$$

is surjective, if $i \geq n+2$, $j \geq 1$. Since $\mathcal{S}\,(L^i,L) = (0)$, $i \geq n+2$ apply the Proposition to prove that:

$$\mathcal{R}\,(L^i,L^j) \otimes \Gamma(L) \longrightarrow \mathcal{R}\,(L^{i+1},L^j)$$

is surjective, if $i \geq n+2$, $j \geq 1$. Iterating again, we get the required assertion.

QED

§2. The case of curves.

For the whole of this section, X will be assumed to be a non-singular complete curve of genus g. We want to strengthen the results of §1 in this case. We need some more concepts and definitions. A divisor $\mathfrak{A}$ is a formal linear combination $\Sigma n_i x_i$ of points of X. For all divisors $\mathfrak{A}$, $\mathfrak{O}(\mathfrak{A})$ is the invertible sheaf of functions f which are regular except at the x_i's, and at x_i have at most an n_i-fold pole, if $n_i \geq 0$, or must have at least a $(-n_i)$-fold zero if $n_i \leq 0$. A fact that we need is that if an invertible sheaf L has a section s with zeroes exactly at $x_1, \cdots, x_k$ of multiplicities $n_1, \cdots, n_k$, then $L \cong \mathfrak{O}(\mathfrak{A})$, with $\mathfrak{A} = \Sigma n_i x_i$. If L is an invertible sheaf, $L(\mathfrak{A})$ stands for $L \otimes \mathfrak{O}(\mathfrak{A})$. Ω will be the sheaf of regular differentials on X.

Theorem 6: Let L,M be invertible sheaves on X such that $\deg L \geq 2g+1$, $\deg M \geq 2g$. Then $\mathcal{S}(L,M) = (0)$.

Proof: Let $d = \deg L$. $\mathfrak{A}$ is to be a positive divisor of degree $d-(g+1)$ which will be chosen later. Then $L(-\mathfrak{A})$ is naturally a subsheaf of L, and we get an exact sequence:

$$0 \longrightarrow L(-\mathfrak{A}) \longrightarrow L \longrightarrow L^* \longrightarrow 0$$

where $\operatorname{Supp} L^* = \operatorname{Supp} \mathfrak{A}$. The 1st requirement on $\mathfrak{A}$ is that $H^1(L(-\mathfrak{A})) = (0)$. Assuming for the moment that $\mathfrak{A}$ has this property, by the 6-lemma of §1, we get an exact sequence:

$$\mathcal{S}\,(L(-\mathfrak{A}),M) \longrightarrow \mathcal{S}\,(L,M) \longrightarrow \mathcal{S}\,(L^*,M)\,.$$

But it is well known that if K is an invertible sheaf on X with deg $K \geq 2g$, $\Gamma(K)$ has no base points. In particular, $\Gamma(M)$ has no base points, and Supp(L^*) is 0-dimensional. So by Castelnuovo's lemma, $\mathcal{S}\,(L^*,M) = (0)$.

Next, apply the Riemann-Roch theorem to $L(-\mathfrak{A})$:

$$\dim H^0(L(-\mathfrak{A})) = \deg L(-\mathfrak{A}) - (g-1) + \dim H^1(L(-\mathfrak{A}))$$

$$= 2.$$

Thus $\Gamma(L(-\mathfrak{A}))$ is a "pencil" and the 2nd requirement on $\mathfrak{A}$ is that it is base point free. Finally we want to apply Castelnuovo's lemma to deduce that $\mathcal{S}\,(L(-\mathfrak{A}),M) = (0)$. For this we need only that

$$H^1(M \otimes L(-\mathfrak{A})^{-1}) = H^1(M \otimes L^{-1}(\mathfrak{A})) = (0).$$

This is the 3rd requirement on $\mathfrak{A}$ Putting all this together, it will follow that $\mathcal{S}\,(L,M) = (0)$.

Can we find an $\mathfrak{A}$ with these 3 properties? Since $\mathfrak{A}$ consists in $d-(g+1) \geq g$ points all of which can be chosen arbitrarily, it is well known that for a suitable choice of $\mathfrak{A}$, $\mathfrak{o}(\mathfrak{A})$ will be isomorphic to any invertible sheaf K of degree $d-(g+1)$. Now the set of all invertible sheaves K of degree $d-(g+1)$ forms a projective variety J, which is exactly the Jacobian of X except that J does not have any natural base point on it to serve as the origin. It suffices to find a K such that

i) $H^1(L \otimes K^{-1}) = (o)$

ii) for all $x \in X$, $\dim H^o(L \otimes K^{-1}(-x)) = 1$

iii) $H^1(M \otimes L^{-1} \otimes K) = (o)$.

Now if (i) is false, $\dim H^o(L \otimes K^{-1}) > 2$ by Riemann-Roch, hence (ii) will be false for <u>all x</u>! Therefore it is enough to check (ii) and (iii) for all x. But by Riemann-Roch,

$$\dim H^o(L \otimes K^{-1}(-x)) > 1 \iff \dim H^1(L \otimes K^{-1}(-x)) > o$$

$$\iff \dim H^o(\Omega \otimes L^{-1} \otimes K(x)) > o$$

$$\iff \exists\, y_1, \cdots, y_{g-2} \text{ such that } \Omega \otimes L^{-1} \otimes K(x) \cong \mathcal{O}(\Sigma y_i)$$

$$\iff \exists\, y_1, \cdots, y_{g-2} \text{ such that } K \cong \Omega \otimes L^{-1}(x - \Sigma y_i).$$

We have only g-1 variable points here, so the locus of K's not satisfying (ii) has dimension at most g-1. Similarly if $\deg M = e + 2g$, we find:

$$H^1(M \otimes L^{-1} \otimes K) \neq (0) \iff H^0(\Omega \otimes M^{-1} \otimes L \otimes K^{-1}) \neq (0)$$

$$\iff \exists\ y_1,\dots,y_k \text{ where}$$

$$k = \deg(\Omega \otimes M^{-1} \otimes L \otimes K^{-1}) = g-1-e$$

$$\text{such that}$$

$$\Omega \otimes M^{-1} \otimes L \otimes K^{-1} \cong \mathcal{O}_C(\Sigma y_i)$$

$$\iff \exists\ y_1,\dots,y_k \text{ such that}$$

$$K \cong \Omega \otimes M^{-1} \otimes L(-\Sigma y_i).$$

Again there are at most g-1 variable points here, so the locus of K's not satisfying (iii) has dimension at most g-1. Since dim J = g, almost all K's do satisfy (ii) and (iii). Thus an $\mathfrak{U}$ with the required properties exists.

QED

Corollary: If L is an invertible sheaf of degree $\geq$ 2g+1, then L is ample with normal generation.

If the argument in the above proof is traced through, it is not hard to show that it proves the following:

$$\exists\ s_1,s_2 \in \Gamma(L)$$

$$\exists\ \quad t \in \Gamma(M) \qquad \text{such that}$$

$$[ks_1 \otimes_k \Gamma(M) + k.s_2 \otimes_k \Gamma(M) + \Gamma(L) \otimes_k k.t] \longrightarrow \Gamma(L \otimes M)$$

is surjective.

Our argument is essentially the same as the classical argument used to prove that if X is not hyperelliptic, then Ω is normally generated (See Hensel-Landsberg). We can paraphrase this argument in our language as follows:

We begin as before with an exact sequence:

$$0 \longrightarrow \Omega(-\mathfrak{A}) \longrightarrow \Omega \longrightarrow \Omega^* \longrightarrow 0$$

where we now assume that $\mathfrak{A}$ is a positive cycle of degree g-2. In order to apply the 6-lemma, it is not necessary that $H^1(\Omega(-\mathfrak{A})) = (0)$. In fact, it is enough if:

i) $H^1(\Omega(-\mathfrak{A})) \longrightarrow H^1(\Omega)$ is an isomorphism.

This is the 1st requirement on $\mathfrak{A}$. We then deduce as before that

$$\mathcal{S}(\Omega(-\mathfrak{A}),\Omega) \longrightarrow \mathcal{S}(\Omega,\Omega) \longrightarrow \mathcal{S}(\Omega^*,\Omega)$$

is exact. Since $\Gamma(\Omega)$ has no base points, we know that $\mathcal{S}(\Omega^*,\Omega) = (0)$ by Castelnuovo's lemma. By the Riemann-Roch theorem, it follows as before that $\Gamma(\Omega(-\mathfrak{A}))$ is a pencil and our 2nd requirement is that it is base point free. Unfortunately, we cannot apply Castelnuovo's lemma to prove $\mathcal{S}(\Omega(-\mathfrak{A}),\Omega) = (0)$, since $H^1(\Omega \otimes \Omega(-\mathfrak{A})^{-1}) = H^1(\mathfrak{o}(\mathfrak{A}))$ is never (0). We use instead a direct computation of dimensions to prove $\mathcal{S}(\Omega(-\mathfrak{A}),\Omega) = (0)$: Let s_1, s_2 be a basis of $\Gamma(\Omega(-\mathfrak{A}))$. Look at the map:

$$\underbrace{\Gamma(\Omega)s_1 \oplus \Gamma(\Omega)s_2}_{\dim = 2g} \xrightarrow{\alpha} \underbrace{\Gamma(\Omega^2(-\mathfrak{A}))}_{\dim = 2g-1}$$

(The dimension on the right is computed by the Riemann-Roch theorem.) We want α to be surjective. But the kernel will be isomorphic to the spaces of pairs $\omega_1, \omega_2 \in \Gamma(\Omega)$ such that $\omega_1 \otimes s_1 = -\omega_2 \otimes s_2$. Since s_1 and s_2 have no common zeroes, this implies that ω_1 is zero at the zeroes $\mathfrak{b}_2$ of s_2, i.e., $\omega_1 = \eta \otimes s_2$ where $\eta \in \Gamma(\Omega(-\mathfrak{b}_2))$. Then ω_2 is necessarily $-\eta \otimes s_1$, so

$$\mathrm{Ker}(\alpha) \cong \Gamma(\Omega(-\mathfrak{b}_2)).$$

Since $\quad \mathfrak{o}(\mathfrak{b}_2) \cong \Omega(-\mathfrak{A})$, it follows that

$$\begin{aligned} \dim \mathrm{Ker}(\alpha) &= \dim \Gamma(\Omega \otimes \Omega(-\mathfrak{A})^{-1}) \\ &= \dim \Gamma(\mathfrak{o}(\mathfrak{A})) \\ &= \dim H^1(\Omega(-\mathfrak{A})) = 1. \end{aligned}$$

Therefore α is surjective, hence $\mathcal{S}(\Omega,\Omega) = (0)$.

Now let $\Omega(-\mathfrak{A}) = K$. K is a sheaf of degree g, and conversely every sheaf K of degree g such that $\dim \Gamma(K) \geq 2$ has the property $\dim \Gamma(\Omega \otimes K^{-1}) \geq 1$ by Riemann-Roch, hence $\Omega \otimes K^{-1} \cong \mathfrak{o}(\mathfrak{A})$, some $\mathfrak{A}$, hence $K \cong \Omega(-\mathfrak{A})$, some $\mathfrak{A}$. Therefore we have proven:

Theorem 7: If X carries an invertible sheaf K of degree g such that $\Gamma(K)$ is a base point free pencil, then $\mathcal{S}(\Omega,\Omega) = (0)$.

The existence of such a K is not hard to show whenever X is <u>not</u> hyperelliptic, but we omit this. The proof that $\mathcal{S}(\Omega,\Omega^i) = (0)$ if $i \geq 2$, is even easier.

Theorem 6 for the vanishing of $\mathcal{S}$ is definitely the best possible unless further restrictions are placed on L and M. For example, if $L = \Omega(P+Q)$, then although L is ample and $\Gamma(L)$ has no base points, $\phi_L(P) = \phi_L(Q)$, so L is not very ample. Since L^2 is very ample, there must be sections $s \in \Gamma(L^2)$ such that $s(P) = 0$, $s(Q) \neq 0$, hence $s \notin \mathrm{Im}(\Gamma(L) \otimes \Gamma(L))$. Therefore $\mathcal{S}(L,L) \neq (0)$!

We now go on to results about $\mathcal{R}$ for curves. I don't think, unfortunately, that my results here are best possible. I shall prove

<u>Theorem 8</u>: **Let L,M,N be invertible sheaves on X such that** $\deg L \geq 3g+1$, $\deg M$, $\deg N \geq 2g+2$. Then

$$\mathcal{R}(L,M) \otimes \Gamma(N) \longrightarrow \mathcal{R}(L \otimes N, M)$$

is surjective.

From this we deduce immediately:

<u>Corollary</u>: Let L be an invertible sheaf on X such that $\deg L \geq 3g+1$. Then L is normally presented.

<u>Proof of the Theorem</u>: We shall use the following lemma:

<u>Lemma</u>: For all invertible sheaves N on X such that $\deg N \geq 2g+2$ and $\Gamma(N)$ has no base points, there is a decomposition:

$$N = N_1 \otimes \cdots \otimes N_k, \qquad k \geq 2$$

where

(1) $\deg N_i = g+1, \quad 1 \leq i \leq k-1$

$$g+1 \leq \deg N_k \leq 2g+1,$$

(2) $\Gamma(N_i)$ has no base points.

(3) If J_1 (resp. J^*) is the variety of invertible sheaves of degree = degree N_1 (resp. deg N_1 - deg N_2), then for all open sets $U_1 \subset J_1$, $U^* \subset J^*$, we may assume

$$N_1 \in U_1$$

$$N_1 \otimes N_2^{-1} \in U^* .$$

Proof: If $\deg N \leq 2g+1$, then let $k = 1$, $N_1 = N$. Now suppose $\deg N = e + (g+1)$, $g+1 \leq e \leq 2g+1$. Then $k = 2$ and we must decompose $N = N_1 \otimes N_2$, $\deg N_1 = g+1$, $\deg N_2 = e$. Let J_2 be the variety of invertible sheaves of degree = e. Let $V_i \subset J_i$ be the set of invertible sheaves K such that $H^1(K) = (0)$ and $\Gamma(K)$ has no base points. It is well known that V_i is open and non-empty. Consider the maps:

$$f: J_1 \longrightarrow J_2, \quad \text{given by} \quad N_1 \longmapsto N \otimes N_1^{-1}$$

$$g: J_1 \longrightarrow J^*, \quad \text{given by} \quad N_1 \longmapsto N_1^2 \otimes N^{-1} .$$

If identity points are chosen arbitrarily on J_1, J_2, J^*, then all these varieties are canonically the same, and are nothing but the jacobian of X. Then in terms of the group law on the jacobian f becomes a map of the form $x \longmapsto a-x$, and g is of the form $x \longmapsto 2x+b$. Thus both f and g are surjective. In particular, $f^{-1}(V_2)$ and $g^{-1}(U^*)$ are non-empty. Now choose $N_1 \in U_1 \cap V_1 \cap f^{-1}(V_2) \cap g^{-1}(U^*)$, and let $N_2 = N \otimes N_1^{-1}$. Then N_1 and N_2 have all the required properties.

If $k > 2$, the proof is similar, but even simpler.

QED

To prove Theorem 8, begin by decomposing the N in the Theorem by the method of the lemma. It clearly will suffice to prove:

$$\mathcal{R}(L \otimes N_1 \otimes \cdots \otimes N_i, M) \otimes \Gamma(N_{i+1}) \longrightarrow \mathcal{R}(L \otimes N_1 \otimes \cdots \otimes N_{i+1}, M)$$

is surjective, for every i with $0 \leq i \leq k-1$. Checking degrees here, we find that we have reduced the Theorem to:

(A) If $\Gamma(N)$ has no base points, $g+1 \leq \deg N \leq 2g+1$, $\deg M \geq 2g+2$, and $\deg L - \deg N \geq 2g$, then

$$\mathcal{R}(L,M) \otimes \Gamma(N) \longrightarrow \mathcal{R}(L \otimes N, M)$$

is surjective.

We now want to apply the Proposition in §1 to interchange M and N in (A). Since $H^1(L \otimes N^{-1}) = (0)$, hence $\mathcal{S}(L,N) = (0)$, (A) is implied by:

(B) If $\Gamma(N)$ has no base points, $g+1 \leq \deg N \leq 2g+1$, $\deg M \geq 2g+2$, and $\deg L - \deg N \geq 2g$, then

$$\mathcal{R}(L,N) \otimes \Gamma(M) \longrightarrow \mathcal{R}(L \otimes M, N)$$

is surjective.

Now decompose M by the method of the lemma. To prove (B) it will suffice to prove:

(i) $\mathcal{R}(L,N) \otimes \Gamma(M_1) \longrightarrow \mathcal{R}(L \otimes M_1, N)$ surjective

(ii) $\mathcal{R}(L\otimes M_1,N) \otimes \Gamma(M_2) \longrightarrow \mathcal{R}(L \otimes M_1 \otimes M_2, N)$ surjective

---

(k) $\mathcal{R}(L \otimes M_1 \otimes \ldots \otimes M_{k-1}, N) \otimes \Gamma(M_k) \longrightarrow \mathcal{R}(L \otimes M_1 \otimes \cdots \otimes M_k, N)$ surjective.

We want to apply Theorem 4 to prove these facts. Since $\Gamma(N)$ and $\Gamma(M_i)$ are base point free, we need only check:

(i) $H^1(L \otimes N^{-1} \otimes M_1^{-1}) = (0)$

(ii) $H^1(L \otimes N^{-1} \otimes M_1 \otimes M_2^{-1}) = (0)$

(k) $H^1(L \otimes N^{-1} \otimes M_1 \otimes \cdots \otimes M_{k-1} \otimes M_k^{-1}) = (0)$.

Now $\deg(L\otimes N^{-1}\otimes M_1^{-1}) \geq 2g-(g+1) = g-1$, so if M_1 lies in a suitable open subset of the Jacobian, (i) will hold. Secondly, $\deg(L\otimes N^{-1}\otimes M_1\otimes M_2^{-1}) \geq 2g + (g+1)-(2g+1) = g$, so if $M_1 \otimes M_2^{-1}$ lies in a suitable open subset of the Jacobian, (ii) will hold. Since the lemma allows us to choose M_1 and $M_1\otimes M_2^{-1}$ in any open sets, (i) and (ii) can be achieved. As for the rest, if, for instance, $k \geq 3$,

$$\deg(L\otimes N^{-1}\otimes M_1\otimes M_2\otimes M_3^{-1}) \geq 2g + (g+1) + (g+1) - (2g+1) = 2g+1$$

so (iii) is automatic. The same holds for all the rest. Thus (B) is proven, hence (A), hence the Theorem. <u>QED</u>

§3. Abelian varieties: the method of theta-groups.

By definition, an abelian variety is a projective variety with a structure of a group such that $(x,y) \longmapsto x+y$ and $x \longmapsto -x$ are morphisms $X \times X \longrightarrow X$ and $X \longrightarrow X$ respectively. We first recall various basic facts about invertible sheaves on such varieties.

(I.) For every X, there is a 2nd abelian variety $\hat{X}$, called its dual, and an invertible sheaf P on $X \times \hat{X}$, called the Poincaré sheaf such that $P|_{X \times \{0\}} \cong \mathcal{O}_X$, $P|_{\{0\} \times \hat{X}} \cong \mathcal{O}_{\hat{X}}$, which is characterized by the non-degeneracy properties:

(a) If $Z \subset \hat{X}$ is a subscheme such that $P|_{X \times Z} \cong \mathcal{O}_{X \times Z}$, then $Z = \{0\}$ with reduced structure,

(b) If $Z \subset X$ is a subscheme such that $P|_{Z \times \hat{X}} \cong \mathcal{O}_{Z \times \hat{X}}$, then $Z = \{0\}$ with reduced structure.

(II.) If $\mathrm{Pic}(X)$ is the group of all invertible sheaves on X, there is a subgroup $\mathrm{Pic}^0(X)$ characterized by the property:

$$L \in \mathrm{Pic}^0(X) \iff T_x^*L \cong L, \quad \text{all} \quad x \in X$$

where $T_x : X \longrightarrow X$ is the map $T_x(y) = x+y$. For all $a \in \hat{X}$, let $P_a = P|_{X \times \{a\}}$, an invertible sheaf on X. Then for all $a \in \hat{X}$, $P_a \in \mathrm{Pic}^0(X)$, and $a \longmapsto P_a$ defines an isomorphism of groups:

$$\hat{X} \cong \mathrm{Pic}^0(X).$$

(III.) For all invertible sheaves L on X, and x,y $\in$ X,

$$T^*_{x+y}L \otimes L \cong T^*_x L \otimes T^*_y L.$$

Therefore $T^*_x L \otimes L^{-1} \in \mathrm{Pic}^o(X)$ and there is a unique homomorphism $\phi_L: X \longrightarrow \hat{X}$ characterized by:

$$P_{\phi_L(x)} \cong T^*_x L \otimes L^{-1} .$$

(IV.) The Riemann-Roch theorem for abelian varieties asserts: if $L = \mathcal{O}(D)$, D a divisor on X, then

$$\chi(L) = (D^g)/g! = \pm\sqrt{\deg \phi_L}$$

If this number is not 0, L is said to be non-degenerate. Then there is exactly one i, called the index of L, for which $H^i(L) \neq (0)$. In particular, if L is ample, then

$$\chi(L) = \dim \Gamma(L) > 0,$$
$$H^i(L) = (0),\ i \geq 1.$$

These facts are all more or less well known. Detailed proofs can be found, for example, in my book "Abelian Varieties", to be published by Oxford University Press in the series "Tata Institute Studies in Mathematics." We require, in addition, another invariant of invertible sheaves, which I call its theta-group. We treat this group first set-theoretically:

<u>Definition:</u> $\mathcal{G}(L)$ = the set of all pairs (x,ϕ), where $x \in X$ and $\phi: L \longrightarrow T_x^*L$ is an isomorphism.

The group law is given by:

$$(x,\phi) \cdot (y,\psi) = (x+y,\ T_y^*\phi \cdot \psi)$$

$$L \xrightarrow{\ \psi\ } T_y^*L \xrightarrow{\ T_y^*\phi\ } T_{x+y}^*L$$

It is easy to see that if $K(L) = \ker(\phi_L)$, then this groups fits into an exact sequence:

$$1 \longrightarrow k^* \xrightarrow{\ i\ } \mathcal{G}(L) \xrightarrow{\ \pi\ } K(L) \longrightarrow 1.$$

if
$$i(\lambda) = (0,\ \text{mult. by } \lambda),$$
$$\pi(x,\phi) = x.$$

Moreover, $i(k^*)$ commutes with everything in $\mathcal{G}(L)$. If, instead of using invertible sheaves, we spoke of line bundles, $\mathcal{G}(L)$ would be just the group of automorphisms of L that cover translations of X. Or if we use the language of divisors and divisor classes, then:

$\mathcal{G}(\mathcal{O}_X(D))$ = the set of pairs (x,f), $f \in k(X)$, such that

$$T_x^{-1}D = D + (f)$$

$((f)$ = divisor of poles and zeroes of $f)$.

The group law in this version is:

$$(x,f) \quad (y,g) = (x+y,\ T_y^*f \cdot g).$$

This group is well known in one case: if $L \in \mathrm{Pic}^{o}(X)$. In this case, $\phi_L \equiv 0$, so $K(L) = X$ and $\mathcal{G}(L)$ is an extension:

$$1 \longrightarrow k^* \longrightarrow \mathcal{G}(L) \longrightarrow X \longrightarrow 1.$$

Serre has studied this case, and has shown that $\mathcal{G}(L)$ is abelian, has a natural structure of algebraic group itself, and that

$$L \longmapsto \mathcal{G}(L)$$

defines an isomorphism:

$$\mathrm{Pic}^{o}(X) \longrightarrow \mathrm{Ext}^1(X, \mathbb{G}_m).$$

One can describe the non-commutativity of $\mathcal{G}(L)$ conveniently as follows: look at the commutators $xyx^{-1}y^{-1}$. Since $K(L)$ is abelian, $\pi(xyx^{-1}y^{-1}) = 1$, and $xyx^{-1}y^{-1} \in k^*$. Moreover, since $k^* \subset$ center $(\mathcal{G}(L))$, if we alter x or y by an element of k^*, $xyx^{-1}y^{-1}$ does not change. Therefore there is a map:

$$e_L: \ K(L) \times K(L) \longrightarrow k^*$$

$$\text{such that } \ xyx^{-1}y^{-1} = e(\pi x, \pi y), \text{ all } x,y \in \mathcal{G}(L).$$

It is easy to check that e_L is bi-multiplicative and skew-symmetric.

In treating characteristic p, we need more than a set-theoretic group $\mathcal{G}(L)$ we need a full group <u>scheme</u> $\mathcal{G}(L)$. This is defined by asking that the S-valued points of $\mathcal{G}(L)$, for every scheme S/k should be functorially isomorphic to the groups of pairs (x,ϕ), where x is an S-valued point of X, and if T_x: X×S ⟶ X×S is translating by x, then

$$\phi: L \otimes \mathcal{O}_S \longrightarrow T_x^*(L \otimes \mathcal{O}_S)$$

is an isomorphism. It fits into an exact sequence of group schemes:

$$1 \longrightarrow \mathbb{G}_m \xrightarrow{\ i\ } \mathcal{G}(L) \xrightarrow{\ \pi\ } K(L) \longrightarrow 1$$

where π is smooth and surjective, and $\mathbb{G}_m$ is the kernel of ϕ_L in the category of group schemes. For details, see the last § of my book on Abelian Varieties.

The theta-group $\mathcal{G}(L)$ acts in a natural way on the cohomology groups $H^i(L)$. In fact, if $(x,\phi) \in \mathcal{G}(L)$, then define the automorphism of $H^i(L)$:

$$U_{(x,\phi)}: H^i(L) \xrightarrow[\approx]{T_x^*} H^i(T_x^*L) \xleftarrow[\approx]{H^i(\phi)} H^i(L).$$

This gives a representation of $\mathcal{G}(L)$ and it works equally well for group schemes or for ordinary groups.

I propose to divide the rest of this section in half: I shall look first in characteristic 0, where only the set-theoretic $\mathcal{G}(L)$'s are needed, and prove a theorem for these; I will then discuss the extension to characteristic p.

So let char(k) = 0 now. First we need some pure group theory: Let K be a finite abelian group, and let $\mathcal{G}$ be a central extension:

$$1 \longrightarrow k^* \longrightarrow \mathcal{G} \longrightarrow K \longrightarrow 1.$$

Call $\mathcal{G}$ <u>non-degenerate</u> if k^* is exactly the center of $\mathcal{G}$. Then if $\mathcal{G}$ is non-degenerate:

(1) explicitly, $\mathcal{G}$ has the form $\mathcal{G} \cong k^* \times A \times \hat{A}$, where A is a finite abelian group, $\hat{A} = \mathrm{Hom}(A,k^*)$, and multiplication is

$$(\lambda,x,\xi)\cdot(\mu,y,\eta) = (\lambda\mu\eta(x),\ x+y,\ \xi+\eta).$$

(2) $\mathcal{G}$ has a unique irreducible representation V in which k^* acts by its natural character. All such representations are sums of V with itself. For $k^* \times A \times \hat{A}$, this representatic.. can be realized by:

$$V = k\text{-valued functions on } A$$

$$U_{(\lambda,x,\xi)}f\ (y) = \lambda\cdot\xi(y)\cdot f(x+y),\quad \forall\ f \in V.$$

(3) If $H \subset \mathcal{G}$ is an abelian subgroup such that $H \cap k^* = \{1\}$, then we can decompose the irreducible representation V in (2) according to the characters of H:

$$V = \bigoplus_{\lambda \in \hat{H}} V_\lambda .$$

Then each V_λ is non-empty, and if $\mathcal{G}'$ is the centralizer of H in $\mathcal{G}$, then

$$\mathcal{G}'/\{\lambda(x)^{-1}\cdot x \mid x \in H\}$$

acts on V_λ, is again a non-degenerate extension, and V_λ is its irreducible representation.

This is all elementary group theory and is easy enough to prove. (See my paper "On the equations defining abelian varieties", Inv. Math., vol. 1). The key result is:

Theta-structure theorem: If L is non-degenerate of index i, then $\mathcal{G}(L)$ is a non-degenerate extension and $H^i(L)$ is its unique irreducibl representation, with k^* acting naturally.

We now prove in characteristic 0:

Theorem 9: Let L be an ample invertible sheaf on an abelian variety Then for all $\alpha, \beta \in \hat{X}$, all $n,m \geq 4$

$$\mathcal{S}(L^n \otimes P_\alpha, L^m \otimes P_\beta) = (0).$$

Proof: We require the preliminary fact:

Lemma: Let L and M be invertible sheaves on an abelian variety such that $\Gamma(L) \neq (0)$, $\Gamma(M) \neq (0)$, and $L \otimes M$ is ample. Then

$$\sum_{\alpha \in \hat{X}} \Gamma(L \otimes P_\alpha) \otimes \Gamma(M \otimes P_{-\alpha}) \longrightarrow \Gamma(L \otimes M)$$

is surjective.

Proof of lemma: If W is the image, let us show that W is invariant under the action of $\mathcal{G}(L \otimes M)$. Note that if $x \in K(L \otimes M)$, then

$$\phi_L(x) + \phi_M(x) = \phi_{L\otimes M}(x) = 0.$$

Therefore if $\beta = \alpha + \phi_L(x) = \alpha - \phi_M(x)$,

$$
\begin{aligned}
T_x^*(L \otimes P_\alpha) &\cong L \otimes [T_x^* L \otimes L^{-1}] \otimes P_\alpha \\
&\cong L \otimes P_{\phi_L(x)} \otimes P_\alpha \\
&\cong L \otimes P_\beta
\end{aligned}
$$

and

$$
\begin{aligned}
T_x^*(M \otimes P_{-\alpha}) &\cong M \otimes [T_x^* M \otimes M^{-1}] \otimes P_{-\alpha} \\
&\cong M \otimes P_{\phi_M(x)} \otimes P_{-\alpha} \\
&\cong M \otimes P_{-\beta} .
\end{aligned}
$$

Therefore we get a diagram:

$$
\begin{array}{ccccc}
\Gamma(L\otimes P_\alpha) & \otimes & \Gamma(M\otimes P_{-\alpha}) & \longrightarrow & \Gamma(L\otimes M) \\
\downarrow T_x^* & & \downarrow T_x^* & & \Big| T_x^* \\
\Gamma(T_x^*(L\otimes P_\alpha)) & \otimes & \Gamma(T_x^*(M\otimes P_{-\alpha})) & \longrightarrow & \Gamma(T_x^*(L\otimes M)) \\
\wr\Vert & & \wr\Vert & & \wr\Vert \\
\Gamma(L\otimes P_\beta) & \otimes & \Gamma(M\otimes P_{-\beta}) & \longrightarrow & \Gamma(L\otimes M)
\end{array}
$$

In other words, under the action of an element $(x,\phi) \in \mathcal{G}(L\otimes M)$ on $\Gamma(L\otimes M)$, the image of $\Gamma(L\otimes P_\alpha) \otimes \Gamma(M\otimes P_{-\alpha})$ is taken into the image of $\Gamma(L\otimes P_\beta) \otimes \Gamma(M\otimes P_{-\beta})$. Therefore W is $\mathcal{G}(L\otimes M)$-invariant.

Now since $\Gamma(L\otimes M)$ is $\mathcal{G}(L\otimes M)$-irreducible, either $W = (0)$ or $W = \Gamma(L\otimes M)$. But if $s \in \Gamma(L)$, $s \neq 0$ and $t \in \Gamma(M)$, $t \neq 0$, then $s\otimes t \in \Gamma(L\otimes M)$ is not 0: so $W \neq (0)$. QED

Returning to the theorem, we use the lemma to reduce the proof of the theorem to the special case $n = m = 4$. In fact, consider the diagram:

$$\begin{array}{ccc} \sum_{\gamma\in\hat{X}} \Gamma(L^n\otimes P_\alpha)\otimes\Gamma(L^{m-1}\otimes P_{\beta+\gamma})\otimes\Gamma(L\otimes P_{-\gamma}) & \xrightarrow{a} & \sum_{\gamma\in\hat{X}} \Gamma(L^{n+m-1}\otimes P_{\alpha+\beta+\gamma})\otimes\Gamma(L\otimes P_{-\gamma}) \\ \downarrow b & & \downarrow c \\ \Gamma(L^n\otimes P_\alpha)\otimes\Gamma(L^m\otimes P_\beta) & \xrightarrow[d]{} & \Gamma(L^{n+m}\otimes P_{\alpha+\beta}). \end{array}$$

By the lemma, c is surjective. By induction on n and m, a is surjective. Therefore d is surjective.

Now assume $n = m = 4$. We must show that the map:

$$\tau:\ \Gamma(L^4\otimes P_\alpha)\otimes\Gamma(L^4\otimes P_\beta) \longrightarrow \Gamma(L^8\otimes P_{\alpha+\beta})$$

is surjective. We need first some simple remarks. One is that if L is any non-degenerate sheaf, then there is a natural isomorphism:

$$\mathcal{G}(L\otimes P_\alpha) \cong \mathcal{G}(L), \qquad \text{all } \alpha\in\hat{X}.$$

In fact, consider the diagram:

$$\begin{array}{ccccccccc} 1 & \longrightarrow & k^* & \longrightarrow & \mathcal{G}(P_\alpha) & \xrightarrow{\pi} & X & \longrightarrow & 1 \\ & & & & \nwarrow^{\rho_\alpha} & & \cup & & \\ & & & & & & K(L) & & \end{array}$$

Since k^* is a divisible group, and $\mathcal{G}(P_\alpha)$ is abelian, it is easy to check that there is a homomorphism ρ_α such that $\pi\cdot\rho_\alpha = id^*$. In othe

---

*If $0\to A\to B\to C\to 0$ is an extension of abelian groups, then it splits whenever A is divisible.

words, the extension $\mathcal{G}(P_\alpha)$ splits over $K(L)$. Then for all $(x,\phi) \in \mathcal{G}(L)$, where $\phi: L \longrightarrow T_x^*L$ is an isomorphism, we get an isomorphism

$$L \otimes P_\alpha \xrightarrow{\phi\otimes\rho_\alpha(x)} T_x^*(L \otimes P_\alpha),$$

hence an element $(x,\phi\otimes\rho_\alpha(x)) \in \mathcal{G}(L \otimes P_\alpha)$.

The second remark is that τ is, in a certain sense, $\mathcal{G}(L^4)$-linear. In fact, define $\delta: \mathcal{G}(L^4) \longrightarrow \mathcal{G}(L^8)$ by

$$\delta(x,\phi) = (x,\phi^{\otimes 2})$$

where $\phi^{\otimes 2}: L^8 \longrightarrow T_x^*L^8$ is just $\phi \otimes \phi$.

Note that δ fits into a diagram:

$$\begin{array}{ccccccccc}
1 & \longrightarrow & k^* & \longrightarrow & \mathcal{G}(L^4) & \longrightarrow & K(L^4) & \longrightarrow & 1 \\
 & & \downarrow{\scriptstyle \lambda\mapsto\lambda^2} & & \downarrow{\scriptstyle \delta} & & \cap & & \\
1 & \longrightarrow & k^* & \longrightarrow & \mathcal{G}(L^8) & \longrightarrow & K(L^8) & \longrightarrow & 1 \\
 & & & & & & \cap & & \\
 & & & & & & X & &
\end{array}$$

Now choose splittings:

$$\rho_\alpha: K(L^8) \longrightarrow \mathcal{G}(P_\alpha)$$
$$\rho_\beta: K(L^8) \longrightarrow \mathcal{G}(P_\beta)$$

and let ρ_α, ρ_β induce a 3rd splitting:

$$\rho_{\alpha+\beta}: K(L^8) \longrightarrow \mathcal{G}(P_{\alpha+\beta}).$$

Use $\rho_\alpha, \rho_\beta, \rho_{\alpha+\beta}$ to define isomorphisms $\mathcal{G}(L^4) \cong \mathcal{G}(L^4 \otimes P_\alpha) \cong \mathcal{G}(L^4 \otimes P_\beta)$

and $\mathcal{G}(L^8) \cong \mathcal{G}(L^8 \otimes P_{\alpha+\beta})$. Then it is immediate that via δ, τ is $\mathcal{G}(L^4)$-linear.

The next step is to split $\mathcal{G}(L^4)$ over X_2, the group of points of X of order 2:

$$\begin{array}{ccccccccc} 1 & \longrightarrow & k^* & \longrightarrow & \mathcal{G}(L^4) & \xrightarrow{\ \pi\ } & K(L^4) & \longrightarrow & 1 \\ & & & & & \nwarrow{\scriptstyle \rho} & \cup & & \\ & & & & & & X_2 & & \end{array}$$

As in the case of $\mathcal{G}(P_\alpha)$, this is possible if we check that the subgroup $\pi^{-1}(X_2)$ is abelian. But $K(L^4) = \mathrm{Ker}(\phi_{L^4}) = \mathrm{Ker}(4\cdot\phi_L)$, so $x \in K(L^4)$ if and only if $4x \in K(L)$. In particular, $X_4 \subset K(L^4)$. Therefore, if $x_1, x_2 \in X_2$ and $x_2 = 2y_2$, $y_2 \in X_4$, and

$$\begin{aligned} e_{L^4}(x_1,x_2) &= e_{L^4}(x_1,2y_2) \\ &= e_{L^4}(2x_1,y_2) \\ &= e_{L^4}(0,y_2) = 1. \end{aligned}$$

Thus $\pi^{-1}(X_2)$ is abelian and ρ exists. We may now decompose all 3 vector spaces under the action of the abelian group $\rho(X_2)$:

$$\Gamma(L^4 \otimes P_\alpha) = \bigoplus_{\ell \in \hat{X}_2} E_\ell$$

$$\Gamma(L^4 \otimes P_\beta) = \bigoplus_{\ell \in \hat{X}_2} F_\ell$$

$$\Gamma(L^8 \otimes P_{\alpha+\beta}) = \bigoplus_{\ell \in \hat{X}_2} G_\ell$$

Note that $\tau(E_\ell \otimes F_m) \subset G_{\ell+m}$, since τ is, in particular, X_2-linear.

Next, I claim that in $\mathcal{G}(L^8)$, $\delta(\mathcal{G}(L^4))$ is the centralizer of $\delta(\rho(X_2))$. Since $\delta(\mathcal{G}(L^4))$ is exactly the inverse image $\pi^{-1}(K(L^4))$ in $\mathcal{G}(L^8)$, and since e_{L^8} computes the commutators in $\mathcal{G}(L^8)$, this is equivalent to saying: $\forall\, x \in K(L^8)$

(*) $\qquad x \in K(L^4) \Longleftrightarrow e_{L^8}(x,y) = 1 \qquad\qquad {}_2$.

But if $y \in X$, then $y \in K(L^8) \Longleftrightarrow 2y \in K(L^4)$. Since X is divisible, $K(L^4) = 2\cdot K(L^8)$. Therefore, if we abbreviate $K(L^8) = K$, (*) comes down to the assertion:

(**) $\quad \forall\, x \in K, \quad x \in 2K \Longleftrightarrow e(x,y) = 1$, all $y \in K$ such that $2y = 0$.

Since $\mathcal{G}(L^8)$ is a non-degenerate extension, e is a non-degenerate skew-symmetric form on K, and (**) is clearly true.

We can now apply the 3rd set of statements about non-degenerate extensions that we listed above. We deduce:

1) that each E_ℓ, F_ℓ, G_ℓ is non-empty,

2) that G_ℓ is an irreducible $\delta(\mathcal{G}(L^4))$-module.

The theorem now follows. By (1), choose $s \in E_\ell$, $t \in F_m$ with $s \neq 0$, $t \neq 0$. Then $\tau(s \otimes t)$ is the section $s \otimes t$ of $L^8 \otimes P_{\alpha+\beta}$, which is not zero. So the image of τ contains at least one non-zero element of G_ℓ, for each ℓ. But the image of τ is invariant under $\delta(\mathcal{G}(L^4))$, so by (2), it contains all of G_ℓ. Thus τ is surjective.

<u>QED</u>

Now consider the case char(k) = p ≠ 0. To make the proof work we must use the full group scheme $\mathcal{G}(L)$. First we need some theory about group schemes $\mathcal{G}$ which are central extensions of the type:

$$1 \longrightarrow \mathbb{G}_m \longrightarrow \mathcal{G} \longrightarrow K \longrightarrow 1$$

where K is a finite commutative group scheme. As before, we call $\mathcal{G}$ <u>non-degenerate</u> if $\mathbb{G}_m$ is the full scheme-theoretic center of $\mathcal{G}$ (i.e., ∀ S-valued points x of $\mathcal{G}$, if x commutes with all S'-valued points y of $\mathcal{G}$ for all S'/S, then x should be a point of $\mathbb{G}_m$). There is no simple structure theorem for such $\mathcal{G}$'s. However, they do satisfy

(2') $\mathcal{G}$ has a unique irreducible representation V in which $\mathbb{G}_m$ acts by its natural character. All such representations are sums of V with itself.

(3') If $H \subset \mathcal{G}$ is an abelian subgroupscheme such that $H \cap \mathbb{G}_m = \{1\}$ scheme-theoretically, and if $R_H = \Gamma(\mathcal{O}_H)$ regarded as a representation of H (the "regular representation"), then $V \cong R_H^m$ for some m as an H-space. In particular, for all characters $\lambda: H \longrightarrow \mathbb{G}_m$, the eigenspace $V_\lambda \subset V$ for λ is non-empty. Moreover, if $\mathcal{G}'$ is the scheme-theoretic centralizer of H in $\mathcal{G}$, then

$$\mathcal{G}'/\{\lambda(x)^{-1}\cdot x \mid x \in H\}$$

acts on V_λ, is again a non-degenerate extension, and V_λ is its irreducible representation.

Note that in (3') $V \supset \oplus V_\lambda$, but if $\mathrm{char}(k) \mid \mathrm{order}(H)$, it is possible that $V \underset{\neq}{\supset} \oplus V_\lambda$. In compensation, we have the extra fact, $V \cong R_H^m$.

Next, we still have:

<u>Theta-structure theorem</u>: If L is non-degenerate of index i, then $\mathcal{G}(L)$ is a non-degenerate extension and $H^i(L)$ is its unique irreducible representation, with $\mathbb{G}_m$ acting naturally.

The proofs of these facts are, unfortunately, not yet published. Now let's generalize the proof of Theorem 9 to char(p).

(I.) The lemma remains true. However to prove it, it is necessary to show that for all rings R/k, all R-valued points α of $\mathcal{G}(L \otimes M)$, the automorphism of the R-module $\Gamma(L\otimes M) \otimes_k R$ induced by α takes $W \otimes R$ into itself. This follows as before provided that we first prove the following:

(*) For all R-valued points α of $\hat{X}$, if P_α is the invertible sheaf $(1\times\alpha)^*P$ on $X \times \mathrm{Spec}(R)$, then the image of the map:

$$\Gamma(p_1^*L\otimes P_\alpha) \otimes \Gamma(p_1^*M\otimes P_{-\alpha}) \longrightarrow \Gamma(p_1^*(L\otimes M)) = \Gamma(L\otimes M)\otimes_k R$$

is contained in $W \otimes_k R$.

First if R is a finitely generated integral domain over k, then the intersection of the maximal ideals in R is (0): so to prove that an element $x \in \Gamma(L\otimes M) \otimes_k R$ is in $W \otimes_k R$ for such an R, it suffices to show that for all homomorphisms $\phi: R \longrightarrow k$, the image $1 \otimes \phi(x) \in \Gamma(L \otimes M)$ is in W. And this is just a case of (*) for a k-valued point of X, i.e., it is part of the hypothesis. But since X is an integral scheme of finite type over k, for any R, and any R-valued point α of X, α is induced by an R'-valued point β of X via a homomorphism $R' \longrightarrow R$, with R' an integral domain finitely generated over k. And if (*) is true for β, it follows immediately for α. This proves (*) in general.

(II.) Once the lemma is proven, Theorem 9 is reduced to the case $n = m = 4$ exactly as before.

(III.) Next, isomorphisms $\mathcal{G}(L) \cong \mathcal{G}(L \otimes P_\alpha)$, $\alpha \in \hat{X}$, L non-degenerate, can be set up exactly as before. We need only the well-known lemma:

<u>Lemma</u>: If $0 \longrightarrow \mathbb{G}_m \longrightarrow \mathcal{G} \longrightarrow K \longrightarrow 0$ is an abelian extension, and K is a finite group scheme, then $\mathcal{G} \cong \mathbb{G}_m \times K$.

Moreover, we get a homomorphism of group schemes $\delta: \mathcal{G}(L^4) \longrightarrow \mathcal{G}(L^8)$ exactly as before, and τ turns out again to be $\mathcal{G}(L^4)$-linear.

(IV.) Now, if $\text{char}(k) \neq 2$, the rest of the proof works over k without alteration: $\mathcal{G}(L^4)$ splits over X_2, the vector spaces $\Gamma(L^4 \otimes P_\alpha)$, $\Gamma(L^4 \otimes P_\beta)$, $\Gamma(L^8 \otimes P_{\alpha+\beta})$ split into eigenspaces, and we apply statement (3') about the group theory of non-degenerate $\mathcal{G}$'s. <u>However</u>, if char(k) = 2, X_2, the kernel of multiplication by 2, is never a reduced group scheme. We can still split $\mathcal{G}(L^4)$ over X_2, and $\delta(\mathcal{G}(L^4))$ is still the centralizer of $\delta(\rho(X_2))$ in $\mathcal{G}(L^8)$, <u>but</u> since the representations of X_2 are not completely reducible,

$$\Gamma(L^4 \otimes P_\alpha) \supsetneq \bigoplus_{\ell \in \hat{X}_2} E_\ell \;, \quad \text{etc.}$$

We must finish the proof in a new way. Let W = image of τ. Let $W^\perp \subset \Gamma(L^8 \otimes P_{\alpha+\beta})^*$ be the space of linear maps that kill W. Assume $W \subsetneq \Gamma(L^8 \otimes P_{\alpha+\beta})$, hence $W^\perp \neq (0)$ Now W and hence $W^\perp$ is invariant under the action of $\mathcal{G}(L^4)$, hence of the action of $\rho(X_2)$. Therefore $W^\perp$ contains an eigenvector for at least one character $\ell \in \hat{X}_2$. Let $G_\ell^* \subset \Gamma(L^8 \otimes P_{\alpha+\beta})^*$ be the eigenspace for the character ℓ. Now $\Gamma(L^8 \otimes P_{\alpha+\beta})^*$ is an irreducible representation space for the opposed group to $\mathcal{G}(L^8)$, i.e., with multiplication reversed, and in this representation $\mathbb{G}_m$ acts by its natural character. Therefore applying statement (3') to this opposed group, it follows that G_ℓ^* is $\mathcal{G}(L^4)$-irreducible. Therefore $W^\perp \supset G_\ell^*$.

Now we must construct something inside W. By (3') for $\mathcal{G}(L^4)$, $\Gamma(L^4 \otimes P_\beta)$ contains a non-zero $\rho(X_2)$-invariant t. For all $s \in \Gamma(L^4 \otimes P_\alpha)$, $s \neq 0$, the element $\tau(s \otimes t) \in \Gamma(L^8 \otimes P_{\alpha+\beta})$ is not zero, so τ defines an isomorphism of $\Gamma(L^4 \otimes P_\alpha) \otimes s$ with a subspace $W_o \subset W$. As a representation space for $\rho(X_2)$, W_o is therefore isomorphic to $\Gamma(L^4 \otimes P_\alpha)$, hence to R_2^m, where R_2 denotes the regular representation of X_2. Since R_2 is an injective object in the category of representations of X_2, it follows that:

$$\Gamma(L^8 \otimes P_{\alpha+\beta}) \cong W_o \oplus \widetilde{W}$$

where $\widetilde{W}$ is also X_2-invariant. Now the dual space to R_2 contains eigenvectors for every character of X_2: so there is an element $x \in W_o^*$ which is an eigenvector for the character ℓ. Extend x to a linear map on $\Gamma(L^8 \otimes P_{\alpha+\beta})$ that is zero on $\widetilde{W}$. Then $x \in G_\ell^*$, but $x \not\equiv 0$ on W, i.e., $x \notin W^\perp$. This is a contradiction.

<u>QED</u>

§-. Abelian varieties: the method of the variable pencil.

First of all, we need some more results about the index of invertible sheaves. For proofs of these results, see my book on Abelian Varieties and the appendix to this paper by George Kempf.

Definition: Let L be a degenerate invertible sheaf on an abelian variety X. Let $K = K(L)^o$, the connected component of $K(L)$, $Y = X/K$, and $\pi: X \longrightarrow Y$ the canonical map. Then there is a non-degenerate sheaf M on Y such that $L \cong P_\alpha \otimes \pi^*M$, some $\alpha \in \hat{X}$ (cf. appendix). We define index (L) to be the interval:

$$[\text{index } (M), \text{ index } (M) + \dim K].$$

The following result is proven in the appendix:

Proposition: If $i \notin$ index (L), then $H^i(L) = (o)$.

Now suppose L and M are 2 invertible sheaves on X, and L is ample. Consider the collection of sheaves $L^p \otimes M^q$ and the polynomial:

$$P(p,q) = \chi(L^p \otimes M^q).$$

The following theorem is proven in §16 of my book and in the appendix to this paper:

Theorem: If $g = \dim X$, then there are $\alpha_1, \cdots, \alpha_g \in \mathbb{R}$, $\alpha_1 \geq \alpha_2 \geq \cdots \geq \alpha_g$ such that

$$P(x,y) = \prod_{i=1}^{g} (x - \alpha_i y).$$

Moreover, for all $p, q \in \mathbb{Z}$, $q > 0$, if $\alpha_{i+1} = \cdots = \alpha_{i+k} = \frac{p}{q}$,

$\alpha_i > \frac{p}{q} > \alpha_{i+k+1}$, then

$$\text{index } (L^p \otimes M^q) = [i, i+k].$$

The precise result that we need is slightly stronger. I want to assume only that $\Gamma(L) \neq (0)$ (i.e., $L = \pi^* L_0$ for some $\pi: X \longrightarrow X/K$, L_0 ample on X/K). In this case, I claim:

Theorem: Suppose L and M are 2 invertible sheaves, $\Gamma(L) \neq (0)$ and M non-degenerate. Then

$$P(x,y) = \prod_{i=1}^{r} (x - \alpha_i y) \cdot y^{g-r}$$

for some r and some $\alpha_i \in \mathbb{R}$ with $\alpha_1 \geq \cdots \geq \alpha_r$. For $N \gg 0$, let $i_0 = \text{index } (L^N \otimes M)$. Then for all p,q, $q > 0$, if $\alpha_{i+1} = \cdots = \alpha_{i+k} = p/q$, $\alpha_i > \frac{p}{q} > \alpha_{i+k+1}$, then

$$\text{index } (L^p \otimes M^q) = [i + i_0, i + k + i_0].$$

This theorem is deduced easily from the 1st one, by introducing an ample L_1 and considering all the sheaves $L^p \otimes L_1^{p_1} \otimes M^q$ and the polynomial: $P(p,p_1,q) = \chi(L^p \otimes L_1^{p_1} \otimes M^q)$. We omit this step.

The purpose of this section is to prove:

Theorem 10: Let X be an abelian variety, L an ample invertible sheaf and $n \geq 4$ an integer. L^n defines an immersion:

$$\phi_{L^n}: X \longrightarrow \mathbb{P}(\Gamma(L^n)) .$$

Then $\phi_{L^n}(X)$ is ideal-theoretically an intersection of quadrics of rank ≤ 4.

Proof: First, let's construct a set of quadrics containing $\phi_{L^n}(X)$. Once and for all, fix p and q, $p,q \geq 2$, such that $n = p+q$. Consider the map:

$$\Gamma(L^p \otimes P_\alpha) \otimes \Gamma(L^q \otimes P_{-\alpha}) \longrightarrow \Gamma(L^n).$$

If $s \in \Gamma(L^p \otimes P_\alpha)$, $t \in \Gamma(L^q \otimes P_{-\alpha})$, let the induced section $s \otimes t$ of $\Gamma(L^n)$ be denoted $\langle s,t\rangle$ to prevent a confusion of notation. Then for all $s_1, s_2 \in \Gamma(L^p \otimes P_\alpha)$, $t_1,t_2 \in \Gamma(L^q \otimes P_{-\alpha})$, we get 4 sections of L^n: $\langle s_i, t_j\rangle$, $i,j = 1$ and 2. In $\Gamma(L^{2n})$, we get an identity:

$$\langle s_1,t_1\rangle \otimes \langle s_2,t_2\rangle = \langle s_1,t_2\rangle \otimes \langle s_2,t_1\rangle.$$

Therefore

$$q_{s_1,t_1,s_2,t_2} = \langle s_1,t_1\rangle \otimes \langle s_2,t_2\rangle - \langle s_1,t_2\rangle \otimes \langle s_2,t_1\rangle \in \mathcal{R}(L^n,L^n).$$

If Q_{s_1,t_1,s_2,t_2} is the quadric in $\mathbb{P}(\Gamma(L^n))$ defining by $q_{s_1,t_1,s_2,t_2} = 0$, then we will actually prove:

$$(*)\quad \left\{ \begin{array}{l} \phi_{L^n}(X) \text{ is the ideal-theoretic intersection of the quadrics} \\ \qquad Q_{s_1,t_1,s_2,t_2} \text{ for all } \alpha, s_i, t_i. \end{array} \right.$$

For most of this proof, we will deal with the fact that $\phi_{L^n}(X)$ is the set-theoretic intersection of these quadrics. At the end, we will indicate the easy extension of the method to proof that $\phi_{L^n}(X)$ is also an ideal-theoretic intersection.

The first step is to translate (*) into an assertion on X itself, not involving $\mathbb{P}(\Gamma(L^n))$. The points of $\mathbb{P}(\Gamma(L^n))$ correspond to non-zero linear maps $\ell: \Gamma(L^n) \longrightarrow k$, modulo scalars. Fix one such ℓ. Then it is easy to see that the point defined by ℓ lies on Q_{s_1,t_1,s_2,t_2} if and only if

$$\ell(\langle s_1,t_1\rangle)\cdot\ell(\langle s_2,t_2\rangle) = \ell(\langle s_1,t_2\rangle)\cdot\ell(\langle s_2,t_1\rangle).$$

Moreover, it is elementary linear algebra that this holds for all s_1,t_1,s_2,t_2 if and only if there are linear maps:

$$m_\alpha: \Gamma(L^p\otimes P_\alpha) \longrightarrow k$$
$$n_\alpha: \Gamma(L^q\otimes P_{-\alpha}) \longrightarrow k$$

such that:

$$\ell(\langle s,t\rangle) = m_\alpha(s)\cdot n_\alpha(t), \qquad \text{all} \quad s \in \Gamma(L^p\otimes P_\alpha), \quad t \in \Gamma(L^q\otimes P_{-\alpha}).$$

On the other hand, what does it mean to say that the "point" ℓ is in $\phi_{L^n}(X)$? This means that there is a point $x \in X$, and an isomorphism $L^n \cong \mathcal{O}_X$ near x, such that, evaluating sections by this isomorphism:

$$\ell(s) = s(x), \quad \text{all } s \in \Gamma(L^n).$$

Thus (*) comes down to the assertion:

(**) If $\ell: \Gamma(L^n) \longrightarrow k$ is a non-zero linear map such that for all $\alpha \in \hat{X}$, there exist linear maps $m_\alpha: \Gamma(L^p\otimes P_\alpha)\longrightarrow k$, $n_\alpha: \Gamma(L^q\otimes P_{-\alpha})\longrightarrow k$ for which $\ell(\langle s,t\rangle) = m_\alpha(s).n_\alpha(t)$, then for some $x \in X$, $\ell(s) = s(x)$ all $s \in \Gamma(L^n)$.

In order to prove (**), the basic idea is to treat <u>all α simultaneously</u>, i.e., to put the m_α's and n_α's together into a single homomorphism. In fact, consider the invertible sheaves:

$$p_1^*L^p \otimes P \quad \text{and} \quad p_1^*L^q \otimes P^{-1} \quad \text{on} \quad X \times \hat{X}.$$

These have the property:

$$p_1^*L^p \otimes P\big|_{X\times(\alpha)} \cong L^p \otimes P_\alpha; \qquad p_1^*L^q \otimes P^{-1}\big|_{X\times(\alpha)} \cong L^q \otimes P_{-\alpha}.$$

Define

$$E_p = p_{2,*}(p_1^*L^p \otimes P)$$

$$F_q = p_{2,*}(p_1^*L^q \otimes P^{-1}).$$

Since the higher cohomology groups of $L^p \otimes P_\alpha$, $L^q \otimes P_{-\alpha}$ are zero, E_p and F_q are locally free sheaves on S such that

$$E_p \otimes k(\alpha) \cong \Gamma(L^p \otimes P_\alpha); \quad F_q \otimes k(\alpha) \cong \Gamma(L^q \otimes P_{-\alpha}).$$

There is a natural pairing:

$$E_p \otimes_{\mathcal{O}_{\hat{X}}} F_q \longrightarrow p_{2,*}(p_1^*L^n) \cong \Gamma(L^n) \otimes_k \mathcal{O}_{\hat{X}}.$$

This is the globalized form of the individual pairings $\Gamma(L^p \otimes P_\alpha) \otimes \Gamma(L^q \otimes P_{-\alpha}) \longrightarrow \Gamma(L^n)$. In order to go further, we need:

<u>Lemma 1</u>: If ℓ: $\Gamma(L^n) \longrightarrow k$ satisfies the condition of (**), then for all α, ℓ does <u>not</u> vanish identically on the image of $\Gamma(L^p \otimes P_\alpha) \otimes \Gamma(L^q \otimes P_{-\alpha})$ in $\Gamma(L^n)$.

We will prove the lemma later. Assuming this, we next globalize the m_α and n_α as follows: I claim there is an invertible sheaf K on $\hat{X}$ and surjective homomorphisms:

$$m: E_p \longrightarrow K$$
$$n: F_q \longrightarrow K^{-1}$$

such that the diagram:

$$\begin{array}{ccc} E_p \otimes F_q & \longrightarrow & \Gamma(L^n) \otimes_k \mathcal{O}_{\hat{X}} \\ \downarrow{\scriptstyle m\otimes n} & & \downarrow{\scriptstyle \ell\otimes 1} \\ K \otimes K^{-1} & \xrightarrow{\approx} & \mathcal{O}_{\hat{X}} \end{array}$$

commutes. To see this, consider the composite map:

$$E_p \otimes F_q \longrightarrow \Gamma(L^n) \otimes \mathcal{O}_{\hat{X}} \xrightarrow{\ell\otimes 1} \mathcal{O}_{\hat{X}}.$$

It induces a map of locally free sheaves:

$$m': E_p \longrightarrow \underline{\mathrm{Hom}}\ (F_q, \mathcal{O}_{\hat{X}}) .$$

By the hypothesis in (**), this map, after taking $\otimes k(\alpha)$, is always of rank 0 or 1; by lemma 1, it never has rank 0. Therefore, its image is an invertible subsheaf K of $\underline{\mathrm{Hom}}(F_q, \mathcal{O}_{\hat{X}})$ which is locally a direct summand. m gives a surjective homomorphism $m: E_p \longrightarrow K$. On the other hand, the inclusion of K in $\underline{\mathrm{Hom}}(F_q, \mathcal{O}_{\hat{X}})$ induces a surjection:

$$n: F_q = \underline{\mathrm{Hom}}(\underline{\mathrm{Hom}}(F_q, \mathcal{O}_{\hat{X}}), \mathcal{O}_{\hat{X}}) \longrightarrow \underline{\mathrm{Hom}}(K, \mathcal{O}_{\hat{X}}) = K^{-1}.$$

It is clear that the sheaf K and the homomorphisms m,n make the diagram above commute.

To motivate the next steps, let's imagine that (**) is true and see what K, m, and n ought to turn out to be. For all $x \in X$, let $Q_x = P|_{\{x\}\times\hat{X}}$. Then Q_x is an invertible sheaf on $\hat{X}$ and, if we pick an isomorphism $L^p \xrightarrow{\sim} \mathcal{O}_X$ in a neighborhood of x, then there is a natural restriction map:

$$p_1^* L^p \otimes P \longrightarrow p_1^* L^p \otimes P|_{\{x\}\times\hat{X}} \cong P|_{\{x\}\times\hat{X}}.$$

This induces a map of locally free sheaves on $\hat{X}$:

$$r_x: E_p \longrightarrow Q_x$$

which is a global form of the linear maps:

$$E_p \otimes k(\alpha) \cong \Gamma(L^p \otimes P_\alpha) \xrightarrow{\text{"evaluation at x"}} k.$$

Similarly there is a map:

$$s_x: F_q \longrightarrow Q_x^{-1}$$

which is a global form of the linear maps:

$$F_q \otimes k(\alpha) \cong \Gamma(L^q \otimes P_{-\alpha}) \xrightarrow{\text{"evaluation at x"}} k.$$

Therefore, what we want to prove is:

$$(***) \quad \begin{cases} K \cong Q_x, \text{ for some } x \in X \\ \text{and } m \text{ is a multiple of } r_x, \quad n \text{ of } s_x. \end{cases}$$

If we prove (***), then it follows immediately that ℓ, as a point of $\mathbb{P}(\Gamma(L^n))$, equals $\phi_{L^n}(x)$. In fact, choosing an isomorphism of L^n and $\mathcal{O}_X$ near x, let $\ell': \Gamma(L^n) \longrightarrow k$ by the evaluation map $s \longmapsto s(x)$.

Then what (***) asserts is that the 2 composite homomorphisms:

$$\begin{matrix} L \\ L'' \end{matrix} : E_p \otimes F_q \longrightarrow \Gamma(L^n) \underset{k}{\otimes} \mathcal{O}_{\hat{X}} \underset{\ell'\otimes 1}{\overset{\ell\otimes 1}{\rightrightarrows}} \mathcal{O}_{\hat{X}}$$

differ by a scalar. Say $L = \lambda.L'$. Then on the image of each map

$$\Gamma(L^p \otimes P_\alpha) \otimes \Gamma(L^q \otimes P_{-\alpha}) \longrightarrow \Gamma(L^n),$$

$\ell = \lambda\cdot\ell'$. By the lemma of §3, these images generate $\Gamma(L^n)$, so $\ell = \lambda.\ell'$ on all of $\Gamma(L^n)$ and the Theorem is proven.

To prove (***), we proceed as follows. First apply Serre duality to the morphism $p_2: X\times\hat{X} \longrightarrow \hat{X}$:

$$\begin{aligned} \Gamma(\hat{X}, \underline{\mathrm{Hom}}(E_p, K)) &= \Gamma(\hat{X}, \underline{\mathrm{Hom}}(p_{2,*}(p_1^* L^p \otimes P), K)) \\ &\cong \Gamma(\hat{X}, R^g_{p_{2,*}}(p_1^* L^{-p} \otimes P^{-1} \otimes p_2^* K)) \end{aligned}$$

Since all the cohomology groups of the restriction $p_1^* L^{-p} \otimes P^{-1} \otimes p_2^* K|_{X\times\{\alpha\}} \cong L^{-p} \otimes P_{-\alpha}$ are zero, except for the gth group

$$R^i p_{2,*}(p_1^* L^{-p} \otimes P^{-1} \otimes p_2^* K) = (0), \quad i \neq g.$$

Therefore, we conclude by the Leray spectral sequence that:

$$\Gamma(\hat{X}, \underline{\mathrm{Hom}}(E_p, K)) \cong H^g(X\times\hat{X}, p_1^* L^{-p} \otimes P^{-1} \otimes p_2^* K).$$

Similarly:

$$\Gamma(\hat{X}, \underline{\mathrm{Hom}}(F_q, K^{-1})) \cong H^g(X\times\hat{X}, p_1^* L^{-q} \otimes P \otimes p_2^* K^{-1})$$

hence by Serre duality on $X\times\hat{X}$:

$$\Gamma(\hat{X}, \underline{Hom}(F_q, K^{-1}))^* \cong H^g(X \times \hat{X}, p_1^*L^q \otimes P^{-1} \otimes p_2^*K).$$

Therefore, we have at our disposal the 2 apparently meagre bits of information:

$$H^g(X \times \hat{X}, p_1^*L^m \otimes P^{-1} \otimes p_2^*K) \neq (0), \quad \text{for} \quad m = -p \text{ and } q.$$

But, amazingly, these facts turn out to trigger a Rube Goldberg-like set of cohomological implications that we will describe later. We summarize this part of the proof for now in:

Lemma 2: Let L be ample on X, K any invertible sheaf on $\hat{X}$. If there exist integers $a,b \geq 2$ such that

$$H^g(X \times \hat{X}, p_1^*L^m \otimes P^{-1} \otimes p_2^*K) \neq (0)$$

for $m = -a$ and b, then, in fact, for <u>all</u> m:

i) $p_1^*L^m \otimes P^{-1} \otimes p_2^*K$ is non-degenerate of index g,

ii) $\dim H^g(p_1^*L^m \otimes P^{-1} \otimes p_2^*K) = 1$,

iii) $K \in \mathrm{Pic}^o(\hat{X})$.

But by the theorem of biduality, the invertible sheaf P on $X \times \hat{X}$ makes X into the dual $\hat{\hat{X}}$ of $\hat{X}$ with Poincaré sheaf still P. Therefore, all sheaves in $\mathrm{Pic}^o(\hat{X})$ are isomorphic to Q_x, some $x \in X$, hence $K \cong Q_x$, some $x \in X$.

Finally to show that m is a multiple of r_x, and n is a multiple of s_x, it suffices to prove that

$$\dim \Gamma(\hat{X}, \underline{Hom}(E_p, K)) = 1$$
$$\dim \Gamma(\hat{X}, \underline{Hom}(F_q, K^{-1})) = 1.$$

But we saw above that these dimensions equal

$$\dim H^g(X\times\hat{X},\ p_1^*L^{-p}\otimes P^{-1}\otimes p_2^*K)$$

and

$$\dim H^g(X\times\hat{X},\ p_1^*L^q\otimes P^{-1}\otimes p_2^*K)$$

and these are both 1 by lemma 2. This proves (***)!

We now go on to the lemmas:

<u>Proof of lemma 1</u>: Suppose $\ell \equiv 0$ on the image of $\Gamma(L^p\otimes P_\alpha)\otimes\Gamma(L^q\otimes P_{-\alpha})$. Since ℓ is not zero everywhere, and since $\Gamma(L^p\otimes P_\beta)\otimes\Gamma(L^q\otimes P_{-\beta})$ generate $\Gamma(L^n)$ as β varies, choose a point $\gamma\in\hat{X}$ such that

$$\ell \not\equiv 0 \text{ on } \Gamma(L^p\otimes P_{\alpha+\gamma})\otimes\Gamma(L^q\otimes P_{-\alpha-\gamma})$$

By the hypothesis on ℓ, ℓ on this last space is of the form $m\otimes n$, where $m\not\equiv 0$ and $n\not\equiv 0$. By the same reasoning, for <u>almost all</u> $\delta\in\hat{X}$,

$$m\not\equiv 0 \quad\text{on}\quad \Gamma(L^{p-1}\otimes P_{\alpha+\gamma+\delta})\otimes\Gamma(L\otimes P_{-\delta})\ ,$$

and again for <u>almost all</u> $\delta\in\hat{X}$

$$n\not\equiv 0 \quad\text{on}\quad \Gamma(L^{q-1}\otimes P_{-\alpha+\delta})\otimes\Gamma(L\otimes P_{-\gamma-\delta}).$$

Choose a δ for which $m\not\equiv 0$ <u>and</u> $n\not\equiv 0$. Then it follows that $\ell\not\equiv 0$ on the image in $\Gamma(L^n)$ of:

$$[\Gamma(L^{p-1}\otimes P_{\alpha+\gamma+\delta})\otimes\Gamma(L\otimes P_{-\delta})]\otimes[\Gamma(L^{q-1}\otimes P_{-\alpha+\delta})\otimes\Gamma(L\otimes P_{-\gamma-\delta})].$$

But by interchanging the 2nd and 4th factors, this image is the same as the image in $\Gamma(L^n)$ of:

$$[\Gamma(L^{p-1}\otimes P_{\alpha+\gamma+\delta}) \otimes \Gamma(L\otimes P_{-\gamma-\delta})] \otimes [\Gamma(L^{q-1}\otimes P_{-\alpha+\delta}) \otimes \Gamma(L\otimes P_{-\delta})].$$

The map of this 4-way tensor product into $\Gamma(L^n)$ factors through $\Gamma(L^p\otimes P_\alpha) \otimes \Gamma(L^q\otimes P_{-\alpha})$, so this contradicts the assumption that $\ell \equiv 0$ on the image of this space in $\Gamma(L^n)$.

QED

Proof of lemma 2: This is where we will use the theorem quoted in the beginning of this section. First we compute $H^i(X\times\hat{X}, P \otimes p_2^*K^{-1})$. Apply the Leray spectral sequence:

$$H^i(\hat{X}, R^j p_{2,*}(P) \otimes K^{-1}) \Longrightarrow H^{i+j}(X\times\hat{X}, P \otimes p_2^*K^{-1}).$$

But, as is shown in my book, §13:

$$R^j p_{2,*}(P) = (0), \quad i < g$$

$$R^g p_{2,*}(P) = k(0).$$

Therefore:

$$H^{i+g}(X\times\hat{X}, P\otimes p_2^*K^{-1}) = \begin{cases} (0), & i < 0 \\ H^i(\hat{X}, k(0) \otimes K^{-1}), & i \_ 0. \end{cases}$$

Hence $H^i(X\times X, P\otimes p_2^*K^{-1}) = (0)$ if $i \neq g$, and is 1-dimensional if $i = g$.

By Serre duality, the same is true of $P^{-1} \otimes p_2^*K$. Now consider the family of sheaves:

$$M_{p,q} = p_1^*L^p \otimes (P^{-1} \otimes p_2^*K)^q$$

and their Euler characteristics:

$$P(p,q) = \chi(M_{p,q}) .$$

We know by the above computation and by our hypothesis that:

$$(\mathrm{I})\qquad \begin{cases} M_{0,1} \text{ non-degenerate, index} = g \\ g \in \text{index } (M_{b,1}) \\ g \in \text{index } (M_{-a,1}) \end{cases}$$

It follows from the Theorem that $P(x,1)$ has no zeroes in the open interval $-a < x < b$. But now $P(x,1)$ is a real polynomial of x such that

i) P has only real zeroes,

ii) $P(0) = (-1)^g$,

iii) P has no zeroes with $-a < x < b$.

iv) $P(n) \in \mathbb{Z}$, for all $n \in \mathbb{Z}$.

But (i) implies that P has a unique local maximum or minimum between any 2 zeroes: let $-\alpha < 0 < \beta$ $(\alpha, \beta \in \mathbb{R}^+)$ be its zeroes of smallest absolute value. Since $-\alpha < 1 < \beta$, and $|P(1)| \geq 1 = |P(0)|$, P must have a local maximum or minimum between 0 and β; since $-\alpha < -1 < \beta$, and $|P(-1)| \geq 1 = |P(0)|$, P must also have a local maximum or minimum between $-\alpha$ and 0. This is a contradiction — unless P is constant.

Applying the theorem again, it follows that

$$(\mathrm{II})\qquad \begin{cases} M_{p,q} \text{ non-degenerate} \\ \text{index } (M_{p,q}) = g \\ \dim H^g(M_{p,q}) = 1 \end{cases}$$

for all $p,q \in \mathbb{Z}$, $q \neq 0$. This proves (i) and (ii) of the lemma. To prove (iii), apply the Leray spectral sequence:

If $\mathfrak{F}_i = R^i p_{1,*}(P^{-1} \otimes p_2^*K)$, then

$$E_2^{i,j} = H^i(X, L^m \otimes \mathfrak{F}_j) \Longrightarrow H^{i+j}(X \times \hat{X}, p_1^*L^m \otimes P^{-1} \otimes p_2^*K) .$$

In particular, since L is ample, $E_2^{ij} = (0)$ if $i > 0$, $m >> 0$, hence the spectral sequence reduces to :

$$H^o(X, L^m \otimes \mathfrak{F}_j) \cong H^j(X \times \hat{X}, p_1^*L^m \otimes P^{-1} \otimes p_2^*K), \text{ if } m >> 0.$$

Therefore because of (II) the whole sheaf $\mathfrak{F}_j$ must be zero if $j < g$. The spectral sequence now reduces to

$$H^i(X, L^m \otimes \mathfrak{F}_g) \cong H^{i+g}(X \times \hat{X}, p_1^*L^m \otimes P^{-1} \otimes p_2^*K), \text{ all } m ,$$

hence

$$\chi(L^m \otimes \mathfrak{F}_g) = 1, \quad \text{all } m.$$

This shows first that Supp($\mathfrak{F}_g$) is 0-dimensional, since its Hilbert polynomial is a constant; and second, that $\dim H^o(\mathfrak{F}_g) = 1$, hence $\mathfrak{F}_g \cong k(x)$, some $x \in X$.

Now recall from EGA, Ch. 3, §7 that the cohomology of $P^{-1} \otimes p_2^*K$ along the fibre $\{x\} \times \hat{X}$ of p_1 is computed from the higher direct images by a spectral sequence:

$$\mathrm{Tor}_{-i}^{\mathcal{O}_X}(k(x), R^j_{p_{1,*}}(P^{-1} \otimes p_2^*K)) \Longrightarrow H^{i+j}(P^{-1} \otimes p_2^*K|_{\{x\} \times \hat{X}}) .$$

Since $P^{-1} \otimes p_2^*K|_{\{x\}\times\hat{X}} \cong Q_x^{-1} \otimes K$, and since $\mathcal{F}_j = (0)$, $j < g$, we find

$$H^{g-i}(Q_x^{-1} \otimes K) \cong \mathrm{Tor}_i^{\mathcal{O}_X}(k(x),k(x)).$$

Thus $H^i(Q_x^{-1} \otimes K) \neq (0)$, for all i. For $i = 0$, this gives $\Gamma(Q_x^{-1} \otimes K) \neq (0)$, and for $i = g$, this gives (by Serre duality) $\Gamma(Q_x \otimes K^{-1}) \neq (0)$. Therefore $K \cong Q_x$ hence $K \in \mathrm{Pic}^0(\hat{X})$. <u>QED</u>

This completes the proof that $\phi_{L^n}(X)$ is the set-theoretic intersection of the quadrics Q_{s_1,t_1,s_2,t_2}. To prove that it is also ideal-theoretically equal to this intersection, it is enough, as we remarked in the introduction, to prove that for all $x \in X$, the tangent space to $\phi_{L^n}(X)$ at x is the intersection of the tangent spaces to the quadrics Q_{s_1,t_1,s_2,t_2} at x. Equivalently, let $R = k[\epsilon]/(\epsilon^2)$: then we must prove that for all R-valued points x of $\mathbb{P}(\Gamma(L^n))$, x is in $\phi_{L^n}(X)$ if and only if x is in all the quadrics. But such a point x is defined by a k-linear map $\mathcal{l}: \Gamma(L^n) \longrightarrow R$ such that $\mathrm{Image}(\mathcal{l}) \not\subset k\cdot\epsilon$. Translating suitably the conditions that x is in $\phi_{L^n}(X)$ and in the quadrics, we find that the assertion to be proven comes out as:

(**) If ℓ: $\Gamma(L^n) \longrightarrow R$ is a k-linear map with $\mathrm{Im}(\ell) \not\subset k\cdot\epsilon$, such that for all $\alpha \in \hat{X}$, there exist linear maps m_α: $\Gamma(L^p \otimes P_\alpha) \longrightarrow R$ and n_α: $\Gamma(L^q \otimes P_{-\alpha}) \longrightarrow R$ for which $\ell(\langle s,t\rangle) = m_\alpha(s) . n_\alpha(t)$, then for some R-valued point x of X, $\ell(s) = s(x)$, all $s \in \Gamma(L^n)$.

This is proven by a straightforward generalization of our proof for k-valued points. Lemma 1 is unchanged and one finds first an invertible sheaf K on $\hat{X}\times\mathrm{Spec}(R)$ and surjective homomorphisms:

$$m: E_p \otimes_k R \longrightarrow K$$

$$n: F_q \otimes_k R \longrightarrow K^{-1}$$

on $\hat{X} \times \mathrm{Spec}(R)$ which globalize m_α and n_α. For all R-valued points $x: \mathrm{Spec}(R) \longrightarrow X$ of X, define Q_x on $\hat{X}\times\mathrm{Spec}(R)$ to be the pull-back of P by $x \times 1_{\hat{X}}$. We get restriction maps $r_x: E_p \otimes_k R \longrightarrow Q_x$, $s_x: F_q \otimes_k R \longrightarrow Q_x^{-1}$ as before, and (**) reduces as before to:

(***) $K \cong Q_x$, for some R-valued point x of X, and $m = \mu\cdot r_x$, $n = \nu\cdot s_x$ for some units $\mu,\nu \in R$.

But by our proof for k-valued points, we know already that $K|_X \cong Q_{x_o}$ for some k-valued point x_o of X. Therefore, since Pic^o is an "open" subfunctor of Pic, and since X is the dual of $\hat{X}$, it follows immediately that $K \cong Q_x$ for some R-valued point x of X. To prove the rest of (***), it is only necessary to check that

$$\Gamma(\hat{X} \times \mathrm{Spec}(R), \underline{\mathrm{Hom}}\ (E_p, K)) \cong R$$

$$\Gamma(\hat{X} \times \mathrm{Spec}(R), \underline{\mathrm{Hom}}\ (F_q, K^{-1})) \cong R\ .$$

Then since the restriction m_o of m to X is a non-zero multiple of r_{x_o}, m must be a unit times r_x; and similarly for n.

As before, we compute:

$$\Gamma(\hat{X} \times \mathrm{Spec}(R), \underline{\mathrm{Hom}}(E_p, k)) \cong H^g(X \times \hat{X} \times \mathrm{Spec}(R), p_1^* L^{-p} \otimes p_{12}^* P^{-1} \otimes p_{23}^* K)\ .$$

We can then apply the remark:

If L is an invertible sheaf on $Z \times \mathrm{Spec}(R)$ such that $H^i(L|_Z) = (0)$, $i \neq i_o$, then $H^i(L) = (0)$ if $i \neq i_o$ and $H^{i_o}(L)$ is a free R-module.

This completes the proof of Theorem 10.

Appendix*, by George Kempf

Let X be an abelian variety, L an invertible sheaf on X, Y = connected component of $K(L)$, and $p: X \longrightarrow X/Y$ the canonical map.

Theorem 1: (i) If $L|_Y$ is non-trivial, then $H^i(X,L) = (0)$, all i.

(ii) If $L|_Y$ is trivial, there exists a non-degenerate invertible sheaf M on X/Y with $L = p^*M$, and if i_0 = index(M):

$$H^i(X,L) \cong H^{i_0}(X,M) \otimes H^{i-i_0}(Y,\mathcal{O}_Y), \quad \text{all } i.$$

Proof: The theorem follows from:

Lemma 1: $T_x^*L|_Y \cong L|_Y$ for all $x \in X$, and

Lemma 2: Let $P \xrightarrow{f} Z$ be a principal homogeneous space (in the flat topology) with structure group Y, an abelian variety. Then

$$R^i f_*(\mathcal{O}_P) \cong H^i(Y,\mathcal{O}_Y) \otimes \mathcal{O}_Z.$$

By Lemma 1, we see that $L|_Y \in \mathrm{Pic}^0(Y)$ and also that $L|_{x+Y} = L|_{p^{-1}(p(x))}$ is isomorphic to $L|_Y$. Now if $L|_Y$ is non-trivial, then

$$(0) = H^i(Y, L|_Y) = H^i(p^{-1}(p(x)), L|_{p^{-1}(p(x))})$$

for all i (see Mumford, Abelian Varieties, §13). By the theorems on cohomology and base extension, $R^i p_*(L) = (0)$ for all i. The Leray spectral sequence then implies that $H^i(X,L) = (0)$ for all i.

* The results in this appendix were independently discovered by C. P. Ramanujam.

If $L|_Y$ is trivial, hence $L|_{p^{-1}(p(x))}$ is trivial, the see-saw principle shows that if $M = p_*(L)$, then M is an invertible sheaf such that $L = p^*(M)$. This M is clearly non-degenerate. Note that:

$$\begin{aligned} R^i p_*(L) &= R^i p_*(p^*M) \\ &\cong R^i p_*(\mathcal{O}_X) \otimes_{\mathcal{O}_{X/Y}} M \\ &\cong H^i(Y,\mathcal{O}_Y) \otimes_{k} M \qquad \text{(by lemma 2).} \end{aligned}$$

Therefore

$$H^j(X/Y,\ R^i p_*(L)) \cong H^i(Y,\mathcal{O}_Y) \otimes_k H^j(X/Y,\ M),$$

and this is zero unless $j = i_o$, the index of M. Thus the Leray spectral sequence shows:

$$\begin{aligned} H^i(X,L) &\cong H^{i_o}(X/Y,\ R^{i-i_o} p_*(L)) \\ &\cong H^{i-i_o}(Y,\mathcal{O}_Y) \otimes_k H^i(X/Y,\ M). \end{aligned}$$

Proof of Lemma 1: If $m: X\times X \longrightarrow X$ is the addition morphism, we know that $m^*L \otimes p_1^*L^{-1} \otimes p_2^*L^{-1}$ on $X\times X$ is trivial when restricted to $Y\times X$. Define $s: Y \longrightarrow Y\times X$ by $s(y) = (y,x)$. Then

$$\begin{aligned} & s^*(m^*L \otimes p_1^*L^{-1} \otimes p_2^*L^{-1}|_{Y\times X}) \\ &\cong T_x^*L|_Y \otimes L|_Y^{-1} \otimes \mathcal{O}_Y \end{aligned}$$

is also trivial. QED

Proof of lemma 2: Since $P \times_Z P \cong Y \times P$, it will suffice to prove the stronger:

<u>Sublemma</u>: Given $f: X \longrightarrow S$ a morphism of schemes /k such that there exists $\pi: S' \longrightarrow S$ where π is faithfully flat and $\pi_X(\mathcal{O}_{S'}) \cong \mathcal{O}_S$ with the property:

$\exists$ φ,Y and a diagram

$$\begin{array}{ccc} X\underset{S}{\times}S' & \xrightarrow[\sim]{\varphi} & Y\underset{k}{\times}S' \\ f'\big\downarrow & & \big\downarrow p_2 \\ S' & = & S' \end{array}$$

where Y is proper over k. Then we have an isomorphism $R^i f_*(\mathcal{O}_X) \cong H^i(Y, \mathcal{O}_Y) \underset{k}{\otimes} \mathcal{O}_S$.

<u>Proof</u>: $\pi^*(R^i f_*(\mathcal{O}_X)) \cong R^i f'_*(\mathcal{O}_{X\underset{S}{\times}S'})$ since $S' \longrightarrow S$ is a flat base extension and $R^i f'_*(\mathcal{O}_{X\underset{S}{\times}S'}) \cong H^i(Y,O_Y) \underset{k}{\otimes} \mathcal{O}_{S'}$ because of the existence of φ. Because $H^i(Y,\mathcal{O}_Y)$ is finite-dimensional, $R^i f'_*(\mathcal{O}_{X\underset{S}{\times}S'})$ is a vector bundle. Hence $R^i f_*(\mathcal{O}_X)$ is a vector bundle because π is faithfully flat. Now we define an isomorphism

$$\begin{aligned} R^i f_*(\mathcal{O}_X) &\xrightarrow{\sim} \pi_*\pi^* R^i f_*(\mathcal{O}_X) && (\text{since } \pi_*\mathcal{O}_{S'} = \mathcal{O}_S) \\ &\cong \pi_*[H^i(Y,\mathcal{O}_Y)\underset{k}{\otimes} \mathcal{O}_{S'}] \\ &\cong H^i(Y,\mathcal{O}_Y) \otimes_k \mathcal{O}_S && (\qquad " \qquad) \end{aligned}$$

<u>QED</u>

<u>Theorem 2</u>: Let L and M be invertible sheaves on an abelian variety X, with L ample. Let

$$P_{L,M}(n) = \chi(L^n \otimes M).$$

Then (i) all the roots of $P_{L,M}$ are real and dim K(M) is the multiplicity of 0 as a root,

(ii) Counting roots with multiplicities:

$$H^k(X,M) = (0), \quad \text{if } 0 \leq k < \text{number of positive roots}$$

$$H^{g-k}(X,M) = (0), \text{ if } 0 \leq k < \text{number of negative roots.}$$

<u>Proof</u>: The theorem is proven in Mumford, Abelian varieties, §16, for M non-degenerate. It is obvious when $M \in \text{Pic}^0(X)$ because in this case

$$P_{L,M}(n) = \chi(L^n)$$

$$= \frac{(L^g)\cdot n^g}{g!}$$

and X = K(M). Now suppose $X = X_1 \times X_2$, $L = p_1^*L_1 \otimes p_2^*L_2$ and $M = p_1^*M_1 \otimes p_2^*M_2$ where $M_1 \in \text{Pic}^0(X_1)$, M_2 is non-degenerate on X_2 and L_i is ample on X_i. Then by the Künneth formula,

(1) $$P_{L,M}(n) = P_{L_1,M_1}(n)\cdot P_{L_2,M_2}(n)$$

and $K(M) = K(M_1) \times K(M_2)$. So in this case the theorem follows from the above special cases and the Künneth formula.

We shall reduce the theorem to this case. Suppose $f: Y \longrightarrow X$ is an isogeny. Then

(2) $$P_{f^*L,f^*M}(n) = \deg f \cdot P_{L,M}(n)$$

by the Riemann-Roch theorem, and dim $K(f^*M)$ = dim $K(M)$. Therefore assertion (i) is invariant under an isogeny. Let Y be the identity component of $K(M)$ and let Z be a complementary subvariety for Y in X. We have an isogeny $f\colon Y \times Z \longrightarrow X$. Now $Y \subset K(f^*M)$ and if $M_1 = f^*(M)|_Y$, then as in the proof of Theorem 1, $M_1 \in \mathrm{Pic}^0(Y)$ and M is of the form $p_1^*M_1 \otimes p_2^*M_2$ where M_2 is a non-degenerate invertible sheaf on Z. The next problem is to see that the theorem does not depend on the ample L. Then we can replace f^*L by $p_1^*L_1 \otimes p_2^*L_2$ and we have reduced the proof of (i) to a case where (i) has been proven.

<u>Claim: $P_{L,M}$ and $P_{L',M}$ have the same number of positive, zero, and negative roots (counted with multiplicity).</u>

Let δ (resp. δ') be the smallest positive root of $P_{L,M}$ (resp. $P_{L',M}$). Let a (resp. a') be the number of positive roots of $P_{L,M}$ (resp. $P_{L',M}$). Then

a = number of positive roots of $P_{L,M}(t+\epsilon)$, if $0 < \epsilon < \delta$

a' = " " " $P_{L',M}(t+\epsilon')$, if $0 < \epsilon' < \delta'$.

But
$$\begin{aligned} s^g P_{L,M}\left(n + \frac{r}{s}\right) &= P_{L^s,M^s}\left(n + \frac{r}{s}\right) \\ &= \chi(L^{ns+r} \otimes M^s) \\ &= \chi(L^{sn} \otimes (L^r \otimes M^s)) \\ &= P_{L,L^r\otimes M^s}(sn) \\ &= s^g P_{L,L^r\otimes M^s}(n). \end{aligned}$$

So if $0 < \frac{r}{s} < \delta$, then $L^r \otimes M^s$ is non-degenerate and

$$a = \text{number of positive roots of } P_{L,L^r \otimes M^s}$$
$$= \text{index } (L^r \otimes M^s).$$

Now let N be large enough so that $(L')^N \otimes L^{-1}$ is ample and choose r and s so that $0 < r/s < \delta$, $0 < \frac{Nr}{s} < \delta'$. Then

$$a = \text{index } (L^r \otimes M^s)$$
$$\geq \text{index } (((L')^N \otimes L^{-1})^r \otimes L^r \otimes M^s) \text{ (by Th. for non deg.}$$
$$= \text{index } ((L')^{Nr} \otimes M^s)$$
$$= a'.$$

By symmetry, it follows that $a = a'$. The claim is proven similarly for the multiplicity of 0 and the number of negative roots.

To prove (ii), we may assume that $M = p^*N$ for a non-degenerate N on X/Y, since otherwise M has no cohomology at all. We have the commutative diagram:

$$\begin{array}{ccc} Y \times Z & \xrightarrow{f} & X \\ \downarrow & & \downarrow p \\ Z & \xrightarrow{g} & X/Y \end{array}$$

for some isogeny g. Then g^*N is non-degenerate and index (g^*N) = index(N). So:

number of pos. rts of $P_{L,M}$ = number of pos. rts of P_{f^*L,f^*M} (by formula

= number of pos. rts of $P_{L|_Z, g^*N}$ (by formula

= index g^*N (Th in non-deg. case)

= index N.

Now (ii) follows from Theorem 1.

QED

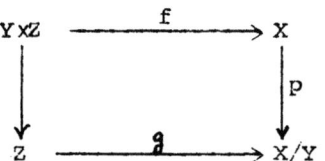

CENTRO INTERNAZIONALE MATEMATICO ESTIVO

(C.I.M.E.)

A. NERON

HAUTEURS ET THEORIE DES INTERSECTIONS

Corso tenuto a Varenna dal 9 al 17 settembre 1969

HAUTEURS ET THÉORIE DES INTERSECTIONS

par

A. Néron

(Université de Orsay)

1 - Introduction - Dans [6] sont étudiés divers problèmes se rattachant à la notion de hauteur d'un point rationnel d'une varieté algébrique définie sur un corps global.

Je me propose ici d'exposer, dans ses grandes lignes, les principaux résultats de ce travail, en commençant par traiter à part, et de façon plus détaillée, le cas particulier des corps de fonctions algébriques d'une variable, mettant en évidence le lien de cette théorie avec la notion de modèle minimal au sens de [5] Bien que les théorèmes fondamentaux du cas général puissent s'obtenir indépendamment de l'existence des modèles minimaux, l'introduction de ceux-ci permet de mieux préciser certains points du cas particulier envisagé, et de présenter la théorie sous une forme purement algébrique, sans recourir aux méthodes de majoration habituelles qui font intervenir la notion "grossière" de hauteur, et sans utiliser aucun passage à la limite.

Les résultats figurant dans cet exposé se trouvent déjà dans [6] Ils sont également résumés dans [7] . Les variantes introduites ne concernent que leur présentation.

2 - Notion de hauteur dans le cas des corps de fonctions algébriques d'une variable

Soit k un corps, et soit K un corps de fonctions algébriques d'une variable sur k. Un tel corps K est de la forme $K = k(t)$, où t est un point générique sur k d'une courbe T, définie sur k, qu'on peut supposer complète et sans point multiple. Pour simplifier l'exposé, on supposera k algébriquement clos. On désignera par $\bar{K}$ une clôture algébrique de K. Ces données sont regardées comme fixes dans la suite, jusquà la fin du n. 6 .

A tout point $a=(a_o, \ldots, a_m)$, rationnnel sur K, de l'espace projectif $\mathbb{P}_m$, on associe un nombre entier $\geqslant 0$, appelé <u>hauteur</u> de a, et noté $h(a)$. Pour le définir, on introduit l'application rationnelle $\lambda_a : T \longrightarrow \mathbb{P}_m$, définie sur k, telle que $\lambda_a(t)= a$, et on considerè un hyperplan H de $\mathbb{P}_m$ tel que le diviseur $\lambda_a^{-1}(H)$ sur T soit défini. La hauteur $h(a)$ est, par définition, le degré de ce diviseur ; on voit immédiatement que ce nombre ne dépend pas du choix de H ; il ne dépend pas non plus du choix de T.

Plus généralement, soit a un point de $\mathbb{P}_m$ algébrique sur K, i.e., rationnel sur $\bar{K}$, et soit $(a_o, \ldots, a_n)$ un système de coordonnées homogènes de a, appartenant à une extension algébrique de degré fini L de K. On peut regarder L comme le corps des fonctions sur une courbe U complète et sans point multiple, définie sur k, i.e. on a $L = k(u)$, où u est un point générique de U sur k. Considérons le k-morphisme $\mu : U \rightarrow T$ tel que $\mu(u) = t$, et posons $n = \deg \mu = [L:K]$. Notons $h_L(a)$ la hauteur de a relative à L. Le nombre $\frac{1}{d} h_L(a)$ ne dépend pas du choix de L ni de celui de U. On le désigne par $h(a)$. Il est clair qu'on obtient là un prolongement de la définition précédente.

Considérons maintenant une sous-variété V de $\mathbb{P}_m$, définie sur K, et soit x un point générique de V sur K. Désignons par $\vec{V}$ la sous-variété de $\mathbb{P}_m \times T$ lieu du point (x,t) sur k, et par μ le morphisme $\bar{V} \to T$ induit par la seconde projection $pr_2 . \mathbb{P}_m \times T \longrightarrow T$. Pour tout point $t^o \in T$, on appelle <u>fibre</u> de t^o le cycle $\mu^{-1}(t^o)$ sur $\bar{V}$. En particulier, la variété V coincide avec la projection sur $\mathbb{P}_m$ de la <u>fibre générique</u> $\mu^{-1}(t)$. Si $\varphi : W \to V$ est une application rationnelle définie sur K, on note $\bar{\varphi}$ l'application rationnelle $\bar{W} \to \bar{V}$, définie sur k, obtenue en prolongeant φ, i.e. telle qu'on ait $\bar{\varphi}(y,t) = (\varphi(y), t)$, pour y générique de W sur K.

On désigne par V_K l'ensemble des points de V qui sont rationnels sur K. Pour $a \in V_K$, le lieu sur k du point (a,t) de $\bar{V}$ est une courbe $\bar{a}$ sur $\bar{V}$, définie sur k. Cette courbe est une <u>section</u> de $\bar{V}$, i.e. est telle que μ induise un k-isomorphisme $\bar{a} \longrightarrow T$; inversement, toute section de $\bar{V}$ est obtenue de cette façon.

On dira qu'une sous-variété de codimension 1 de $\bar{V}$ est <u>verticale</u> si elle est contenue dans l'une des fibres de $\bar{V}$, i.e. si elle coincide avec l'une des composantes de cette fibre. Un diviseur sur $\bar{V}$ dont toutes les composantes sont verticales est dit <u>vertical.</u> Si X est un diviseur sur V, rationnel sur K, il existe un et un seul diviseur $\bar{X}$ sur $\bar{V}$, sans composante verticale, tel que $\bar{X}$ induise X sur la fibre générique.

Pour tout point $t^o \in T$, le sous-ensemble de la fibre V^o de t^o composé des points simples sur cette fibre est noté S_{t^o}, ou $S_{t^o}(V)$. La réunion des ensembles S_{t^o} est un ouvert de $\bar{V}$, qu'on notera $\mathcal{S}(V)$; cet ouvert est lui-même contenu dans celui composé de tous les points simples sur $\bar{V}$.

Si $\varphi : V \to \mathbb{P}_r$ est un K - morphisme de V dans un espace projectif $\mathbb{P}_r$, on note h_φ la fonction sur V_K, à valeurs entières $\geqslant 0$, obtenue en posant $h_\varphi(a) = h(\varphi(a))$ pour tout $a \in V_k$. On note $\mathcal{L}_\varphi$ le système linéaire sur V, défini sur K, attaché à φ (i.e. l'image inverse du système linéaire des hyperplans de $\mathbb{P}_r$), et Γ_φ la classe mudule l'équivalence linéaire, du système $\mathcal{L}_\varphi$.

3 - Variétés abéliennes minimales.

Soit A une variété abélienne définie sur K, partout "faiblement $\mathfrak{p}$-simple $\mathfrak{p}$-minimable" au sens de [5] ; par abréviation, nous dirons simplement que A est "minimale". Dans le cas particulier envisagé, et avec le langage introduit ci-dessus, cette propriété peut être caractérisée comme suit ; pour toute variété V définie sur K, et pour toute K-application rationnelle $\varphi : V \to A$, l'application rationnelle $\overline{\varphi} : \overline{V} \to \overline{A}$ correspondante induit un k-morhisme de $\mathcal{S}(\overline{V})$ dans $\mathcal{S}(\overline{A})$.

Appliquant cette propriété à la loi de groupe sur A, on voit en particulier que, pour tout point $t^o \in T$, l'ensemble $S_{t^o}(A)$ est canoniquement muni d'une structure de groupe algébrique défini sur k ; ce groupe est encore noté $G_{t^o} = G_{t^o}(A)$ De façon plus précise, le schéma sur T déterminé par le morphisme $\mathcal{S}(A) \to T$ est muni canoniquement d'une structure de schéma en groupes, admettant pour fibres les groupes $G_{t^o}(A)$.

La composante neutre de G_{t^o} est notée ${}^oG_{t^o}$. Pour $a \in A_K$, la section $\overline{a}$ correspondante rencontre la fibre A^o de t^o en un point a^o qui appartient nécessairement à

G_{t^o} . En d'autres termes, on a toujours $\bar{a} \subset \mathcal{S}(A)$. Le sous-groupe de A_K composé des points $a \in A_K$ tel que a^o appartienne à la composante neutre ${}^oG_{t^o}$ quel que soit $t^o \in T$ est un sous-groupe d'indice fini de A_K , qu'on note ${}^oA_K$.

Pour tout diviseur X sur A , rationnel sur K , nous noterons $h^*_X = h^*_{X,K}$ l'application $A_K \longrightarrow Z$ définie en posant $h^*_X(a) = \deg(\bar{X}, \bar{a})$, où $\deg(\bar{X}, \bar{a})$ représente le degré global d'intersection de X et $\bar{a}$ sur la variété $\bar{A}$; ce symbole a un sens, puisqu'on a $\bar{a} \subset \mathcal{S}(A)$.

Montrons que si on a $X \sim 0$ (i; e., suivante la convention habituelle, X linéairement équivalent a 0) , la fonction $h^*_X(a)$ ne dépend que de la classe de a (mod ${}^oA_K$) . En effet, il existe une fonction f sur A, définie sur K, telle que $\operatorname{div}(f) = X$. Notons $\bar{f}$ la fonction sur $\bar{A}$, définie sur k, qui prolonge f. Le diviseur $\operatorname{div}(\bar{f}) - \bar{X}$ sur $\bar{V}$ induit le diviseur nul sur la fibre générique, c'est donc un diviseur vertical. Si Z est l'une de ses composantes, le degré $\deg(Z \cdot \bar{a})$ est égal à 1 ou à 0 suivant que $\bar{a}$ rencontre Z ou non; or cette dernière propriété est invariante lorsqu'on modifie a par l'addition d'un élément de ${}^oA_K$, et notre assertion s'en déduit aussitôt .

Il en résulte que si $\varphi : A \longrightarrow \mathbb{P}_r$ est un K-morphisme de A dans un espace projectif, et si $X \in \Gamma_\varphi$, l'expression $h^*_X(a) - h_\varphi(a)$ ne dépend que de la classe de a mod ${}^oA_K$).

Notons aussi qu'on peut prolonger la définition du symbole $h^*_X(a)$ aux points de A qui sont algébriques sur K . Si en effet a est rationnel sur une extension $L \subset \bar{K}$ de degré fini n de K, on peut poser $h^{*}_{X}(a) = \frac{1}{n} h^{*}_{X,L}(a)$; les propriétés classiques du degré d'intersection permettent de vérifier que le second membre ne dépend pas du choix de L.

4. Un lemme sur les fonctions quadratiques.

Soit G un groupe abélien, et soit $f: G \longrightarrow \mathbb{R}$ une fonction sur G, à valeurs réelles. Nous dirons qui f est linéaire si f est un homomorphisme de G dans le groupe additif de $\mathbb{R}$; nous dirons que f est quadratique si f est de la forme $f(x) = F(x, x)$, où $F: G \times G \longrightarrow \mathbb{R}$ est bilinéaire symétrique. La fonction $(x, y) \mapsto f(x+y) - f(x) - f(y)$ sur $G \times G$ est noté $\Delta_1 f$. La fonction

$(x, y, z) \longrightarrow f(x+y+z) - f(y+z) - f(x+z) - f(x+y) + (f(x) + f(y) + f(z)$ sur $G \times G \times G$ est notée $\Delta_2 f$. Pour qu'on ait $\Delta_1 f = 0$ (resp. $\Delta_2 f = 0$), il faut et il suffit que f soit linéaire (resp. soit la somme d'une fonction quadratique et d'une fonction linéaire.

Soit G' un sous-groupe d'indice fini de G. On dira qu'une fonction $f: G \longrightarrow \mathbb{R}$ est périodique relativement à G' si f(a) ne dépend que de la classe de a (mod G'). On dira plus généralement, q'une fonction $g: G \times \dots \times G \longrightarrow \mathbb{R}$ est périodique relativement à G' si $f(a_1, a_2, \dots)$ ne dépend que des classes des a_i (mod G').

Lemme. Soit G un groupe abélien, et soit G' un sous-groupe d'indice fini de G. Soit f une fonction sur G, à valeurs réelles, telle que $\Delta_2 f$ soit périodique relativement à G'. Alors f est de la forme $f = f_0 + \chi$, où f_0 est la somme d'une fonction quadratique et d'une fonction linéaire, et où χ est une fonction périodique relativement à G'. Les fonctions f_0 et χ sont uniquement déterminées par ces conditions. Si f est à valeurs entières, f_0 et χ sont à valeurs rationnelles, admettant un dénominateur fixe.

<u>Démonstration</u> : Si β est une fonction périodique relativement à G', il existe une et une seule fonction α sur G, périodique relativement à G' et telle que $\Delta_2 \alpha = \beta$. On a en effet necessairement

$$\Delta_1 \alpha (x, y) = - \frac{1}{m} \sum_{\mu=1}^{m-1} \beta(x, \mu y)$$

et

$$\alpha (x, y) = - \frac{1}{m} \sum_{\mu=1} \Delta_1 \alpha (x, \mu y)$$

Il existe donc une et une seule fonction réelle χ sur G, périodique relativement à G', et telle que $\Delta_2 \chi = \Delta_2 f$. On a donc $\Delta_2(f - \chi) = 0$, et par suite $f_0 = f - \chi$ est la somme d'une fonction quadratique et d'une fonction linéaire

5 - Le théorème global (cas des corps de fonctions)

<u>Théorème 1</u> : Soit A une varieté abélienne definie sur K et soit X un diviseur sur A, rationnel sur K. Alors il existe une et une seule fonction g_X sur A_K, à valeurs rationnelles,

qui est la somme d'une fonction quadratique et d'une fonction linéaire, et telle que la différence $h^*_X - g_X$ soit périodique relativement au sous-groupe ${}^oA_K$.

La fonction g_X ne dépend que de la classe $\Gamma = \Gamma_X$ de X modulo l'équivalence linéaire, et elle en dépend linéairement.

<u>Autre forme équivalente de l'énoncé</u> : Soit $\varphi : A \to \mathbb{P}_r$ un K-morphisme de A dans un espace projectif. Il existe une et une seule fonction g_φ sur A_K, à valeurs rationnelles, telle que $h^*_\varphi - g_\varphi$ soit périodique relativement à ${}^oA_K$.

Admettant l'énoncé précédent, il suffit en effet compte tenu des remarques faites au n. 3, de prendre $g_\varphi = g_X$, où X est un élément arbitraire de $\mathcal{L}_\varphi$.

<u>Démonstration</u> . Posons $B = A \times A \times A$. Notons π_1, π_2, π_3 les projections respectives $B \to A$ de B sur les trois facteurs. Pour $1 \leqslant i < j \leqslant 3$, posons $\pi_{ij} = \pi_i + \pi_j$ (de sorte qu'on a, par exemple $\pi_{23}(x, y, z) = y+z$. Posons de même $\pi_{123} = \pi_1 + \pi_2 + \pi_3$.

Considérons les diviseurs

$$X_i = \pi_i^{-1}(X) \qquad X_{ij} = \pi_{ij}^{-1}(X) \qquad X_{123} = \pi_{123}^{-1}(X)$$

sur B . On a

$$\Delta_2 h^*_X = h^*_Z \quad ,$$

en posant

$$Z = X_{123} - \sum_{i<j} X_{ij} + \sum_i X_i$$

Or d'après une propriété classique des diviseurs sur les variétés abéliennes (théorème du cube), on a $Z \sim 0$ compte tenu de ce qui précède (n. 3) h^*_Y est une fonction périodique relativement au sous-groupe ${}^{\circ}A_K$.

L'existence de g_X s'en déduit aussitôt, par application du lemme du n. 4. Les autres assertions du théorème sont immédiates.

Théorème 2 . Soient A et B deux variétés abèliennes minimales définies sur K, et soit $\alpha : B \longrightarrow A$ un K-morphisme. Soit X un diviseur sur A, rationnel sur K , et posons $Y = \alpha^{-1}(X)$. Soient b, $b' \in B_K$, et posons $a = \alpha(b)$ et $a' = \alpha(b')$. Alors on a $g_Y(b') - g_Y(b) = g_X(a') - g_X(a)$. En particulier, si α est un K-homomorphisme, on a $g_Y(b) = g_X(a)$.

Démonstration : On a en effet $h^*_X(a) = \deg(\overline{X}.\overline{a}) =$ $= \deg((\overline{\alpha})^{-1}(\overline{Y})\,\overline{b})$. Le diviseur $\overline{Y} - (\overline{\alpha})^{-1}(\overline{X}))$ sur $\overline{B}$ est vertical, donc la fonction $b \mapsto h^*_Y(b) - h^*_X(\alpha(b))$ sur B_K est périodique relativement à ${}^{\circ}B_K$. Le théorème en résulte aussitôt, compte tenu du th. 1, et du fait que tout morphisme $B \longrightarrow A$ est le composé d'un homomorphisme et d'une translation.

Compte tenu du th . 1, et des définitions introduites au n. 3, la définition du symbole $g_X(a)$ se prolonge de façon naturelle au cas où a est un point algébrique sur K , et le th . 2 reste valable lorsqu'on y remplace les points rationnels par des points algébriques sur K .

Soit maintenant $\mathfrak{a}$ un cycle de dimension 0 sur A. Par définition, un tel cycle est une combinaison formelle $\sum_i m_i(a_i)$

de points $a_i \in A$, à coefficients entiers (les parenthèses sont mises pour éviter de confondre l'addition des cycles avec la loi de groupe de A). Si $\mathfrak{a}$ est rationnel sur K, les a_i sont algébriques sur K ; pour un tel $\mathfrak{a}$, on prolonge par linéarité la définition du symbole g_X en posant $g_X(\mathfrak{a}) = $ $= \sum_i m_i \, g_X(a_i)$.

Le symbole $g_X(\mathfrak{a})$, qu'on note encore $(X, \mathfrak{a})$ est, d'après le th. 2, invariant par tout K-isomorphisme pour la structure de variété algébrique (i.e. par le composé d'un isomorphisme pour la structure de variété abélienne et d'une translation).

Compte tenu du théorème d'existence des modèles minimaux ([5], II, 10, th. 4) on peut donc prolonger la définition du symbole $g_X(\mathfrak{a}) = (X, \mathfrak{a})$ au cas d'une variété abélienne A définie sur K, non nécessairement minimale. Il existe en effet un K-modèle minimal (B, ψ) de A ; il suffit de poser $(X, \mathfrak{a}) =$ $= (\psi(X), \psi(\mathfrak{a}))$; ce nombre ne dépend pas du modèle choisi.

Enfin, soit V une variété sans point multiple quelconque, définie sur K ; soit X un élément du groupe $\tilde{D}(V)_K$ (cf. [6], II, 2), i.e. un diviseur X sur V, rationnel sur K, tel qu'il existe un entier m, une variété abélienne B définie sur K, un K-morphisme $\varphi : V \rightarrow B$ et un diviseur Y sur B, rationnel sur K vérifiant la condition $mX \sim \varphi^{-1}(Y)$ (rappelons que le groupe $\tilde{D}(V)$ contient le groupe $D_a(V)$ des diviseurs sur V algébriquement équivalents à 0). Pour tout cycle $\mathfrak{a}$ de dimension et de degré nuls sur A, rationnel sur K, on définit le symbole $(X, \mathfrak{a})$ en posant $(X, \mathfrak{a}) = \frac{1}{m}(Y, \varphi(\mathfrak{a}))$. Utilisant la propriété universelle de la variété d'Albanese de V, on voit sans peine que le second membre de cette relation ne dépend pas de choix de B, φ et Y, que le symbole ainsi défini est invariant par tout K-isomorphisme qu'il est bilinéaire en X et $\mathfrak{a}$ et s'annule lorsque $X \sim 0$.

Pour le calcul explicite de g_X dans certains cas simples, lorsque A est une courbe elliptique, nous renvoyons à [3].

6. Expression de $(X, \mathfrak{a})$ comme somme de termes locaux.

Nous allons montrer qu'on peut exprimer canoniquement le symbole $(X, \mathfrak{a})$ sous forme d'une somme de termes locaux, respectivement associés aux differents points de T. A tout point $t^o \in T$, il correspond canoniquement une valuation discrète $v = v_{t^o}$ de K.

Supposons d'abord que chacun des composants a_i du cycle $\mathfrak{a} = \sum_i m_i (a_i)$ est rationnel sur K. Notons $\bar{\mathfrak{a}}$ le cycle $\sum_i m_i \bar{a}_i$, de dimension 1, sur $\bar{A}$. Notons $i_{t^o}(X, \mathfrak{a})$, ou encore $i_v(X, \mathfrak{a})$, la contribution dans $\deg(\bar{X} \cdot \bar{\mathfrak{a}})$ des points de le fibre A^o de t^o.

Supposons en outre qu'on ait $X \sim 0$, $\deg \mathfrak{a} = 0$, et que X et $\mathfrak{a}$ soient étrangers (i.e. que leurs support soient disjoints). Il existe une fonction f sur V, définie sur K, telle que $\mathrm{div}(f) = X$. Posons $f(\mathfrak{a}) = \prod_i f(a_i)^{m_i}$; puisque $\deg \mathfrak{a} = \sum_i m_i = 0$, l'élément $f(\mathfrak{a})$ de K ne dépend que de X et de $\mathfrak{a}$ mais non de f; on le note encore $X(\mathfrak{a})$. Soit $\bar{f}$ la fonction sur $\bar{A}$, définie sur k, qui prolonge f. Le diviseur $Y = \mathrm{div}(f) - X$ sur $\bar{A}$ est vertical. Désignant par $j_v(X, \mathfrak{a})$

la contribution des points de la fibre A^o dans $\deg(Y', \bar{\mathfrak{a}})$, on a

(1) $$i_v(X, \mathfrak{a}) = v(X(\mathfrak{a})) - j_v(X, \mathfrak{a}) .$$

On pose alors $(X, \mathfrak{a})_v = v(X(\mathfrak{a}))$. Il est clair que chacun des symboles $(X, \mathfrak{a})_v$, $i_v(X, \mathfrak{a})$ et $j_v(X, \mathfrak{a})$ dépend bilinéairement de X et de $\mathfrak{a}$. De plus, l'application $a \longrightarrow j_v(X, (a) - (0))$ de A_K dans $\mathbf{Z}$ est périodique relativement à ${}^oA_K$.

On peut prolonger canoniquement la définition des symboles $(X, \mathfrak{a})_v$ et $j_v(X, \mathfrak{a})$ au cas où X est un diviseur sur A, rationnel sur K, quelconque (non nécessairement ~ 0), le cycle $\mathfrak{a}$ étant toujours de degré 0, et de composants rationnels sur K; les symboles obtenus sont à valeurs rationnelles; ils dépendent bilinéairement de X et de $\mathfrak{a}$, vérifient encore la relation (1), et on peut achever de les caractériser en exigeant les deux conditions suivantes :

(i) - $(X, \mathfrak{a})_v$ et $j_v(X, \mathfrak{a})$ sont invariants par toute translation sur A, rationnelle sur K.

(ii) - L'application $\rho_X = \rho_{X,v} : A_K \longrightarrow \mathbf{Q}$ obtenue en posant $\rho_{X,v}(a) = j_v(X, (a) - (0))$ est périodique relativement à ${}^oA_K$.

La condition (ii) peut encore être remplacée par la condition plus faible suivante

(ii_a) - L'application $\rho_X : A_K \longrightarrow \mathbf{Q}$ est bornée ou bien par la condition (ii) du n. 7 ci-dessous.

Pour prouver l'existence et l'unicité de prolongement, on observe que les conditions ci-dessus entraînent

$$\Delta_2 \rho_{X,v} = \rho_{Z,v}$$

où Z est le diviseur sur $A \times A \times A$ introduit dans la démonstration du th 1 ; on a vu que $Z \sim 0$; il suffit d'appliquer à nouveau le lemme du n. 4 .

On peut ensuite, en remplaçant K par une extension algébrique de degré fini convenable, lever la restriction concernant la rationalité des composants de $\mathfrak{a}$,et définir les symboles $(X, \mathfrak{a})_v$ et $j_v(X, \mathfrak{a})$ lorsque $\mathfrak{a}$ est un cycle arbitraire de dimension et de degré nuls, rationnel sur K, étranger à X. Les propriétés énumérées ci-dessus restent satisfaites.

Enfin, on peut, comme pour le théorème global, passer au cas d'une variété V sans point multiple quelconque, définie sur K, et définir les symboles $(X, \mathfrak{a})_v$ et $j_v(X, \mathfrak{a})$ pour tout diviseur $X \in \widetilde{D}(V)$ et pour tout cycle $\mathfrak{a}$ de dimension et de degré nuls, mutuellement étrangers et rationnels sur K. Les symboles obtenus sont encore invariants par tout K-isomorphisme .

On peut en outre , dans tous les cas, prendre comme dénominateur des nombres $(X, \mathfrak{a})_v$ et $j_v(X, \mathfrak{a})$ un entier qui ne dépend que de K, A et v , mais non de X ni de $\mathfrak{a}$.

7. Cas d'un corps global quelconque.

On désigne maintenant par K un corps global quelconque, c'est-à-dire un corps muni d'une famille M de valeurs absolues propre au sens de [1] , et vérifiant la formule du produit; on écrira celle-ci sous la forme $\sum_{v \in M} v(x) = 0$, utilisant, comme dans [6] , la notation additive.

Dans le cas des corps de fonctions d'une variable, précédemment étudié, on prend pour M la famille des valuations discrètes respectivement associées aux points de la courbe T. L'autre exemple fondamental est celui des corps de nombres (extensions algébriques de degré fini de $\mathbb{Q}$).

Soit $a = (a_o, \ldots, a_m)$ avec $a_i \in K$, un point de l'espace projectif $\mathbb{P}_m$, rationnel sur K . On appelle hauteur de a le nombre réel

$$h(a) = h_K(a) = \sum_{v \in M} \sup_i v(a_i)$$

Ce nombre ne dépend que de a, mais non du choix des a_i, d'après la formule du produit. Plus généralement, soit a un point de $\mathbb{P}_m$ rationnel sur la $\mathbb{P}$ algébrique $\overline{K}$ de K , et soit $(a_o, \ldots, a_m)$ un système de coordonnées de a appartenant à une extension L de degré fini n de K . On appelle hauteur de a le nombre réel

$$h(a) = \frac{1}{n} \sum_w n_{w/v} \sup_i w(a_i)$$

où w parcourt la famille M_L de valeurs absolues de M qui prolonge M, et où $n_{w/v}$ est le degré local de l'extension L/K relativement à w . Le symbole ainsi défini ne dépend pas du choix des coordonnées a_i ni de celui de l'extension L .

Nous dirons que deux fonctions f et g sur un ensemble E, à valeurs réelles, sont équivalentes (ce que nous écrirons $f \approx g$) si la différence f-g est bornée.

Le symbole h_φ , défini comme au n. 2 . possède les deux propriétés suivantes.

<u>Proposition 1</u> . Soient V une variété complète, normale, définie sur K, et soient $\varphi : V \to \mathbb{P}_r$, $\psi : V \to \mathbb{P}_s$ deux morphismes définis sur K . Alors, si $\Gamma_\varphi = \Gamma_\psi$, on $h_\varphi \approx f_\psi$.

En d'autres termes, la classe de h_φ modulo la relation $\approx$ ne dépend que de la classe Γ_φ .

<u>Proposition 2</u>. La classe de h_φ modulo la relation $\approx$ dépend linéairement de Γ_φ (pour les démonstrations, voir [1] IV, 2)

Le théorème global fondamental concernant les hauteurs des points rationnels sur une variété abélienne est le suivant.

<u>Théorème 3.</u> Soit A une variété abélienne définie sur un corps global K , et soit φ un K-morphisme $A \to \mathbb{P}_r$. Il existe une et une seule fonction g_φ sur A_K , à valeurs réelles, qui est la somme d'une fonction quadratique et d'une fonction linéaire, et telle qu'on ait $h_\varphi \approx g_\varphi$. De plus, g_φ ne dépend que de la classe Γ_φ , et en dépend linéairement.

<u>Démonstration</u> (d'après Tate) . On prouve un lemme général , exprimant que, pour toute fonction réelle f sur un groupe abéline G telle que $\Delta_2 f$ soit bornée, il existe g , somme d'une fonction quadratique et d'une fonction linéaire sur G ,

telle que $f \approx g$. Il suffit donc de prouver que $\Delta_2 h_\varphi$ est bornée. Ceci se voit en raisonnant comme dans la démonstration du th. 1 (n. 7) et en tenant compte des prop. 1 et 2 ci-dessus.

Compte tenu du fait que le symbole g_φ ne dépend que de $\Gamma = \Gamma_\varphi$, on le note également g_Γ , ou encore g_X , si X est un élément arbitraire de Γ. Le symbole g_X est ainsi défini pour tout diviseur X ample. On passe ensuite au cas où X est quelconque en utilisant la linéarité de h_X (prop. 2) et le fait que tout diviseur sur A est la différence de deux diviseurs amples. On prolonge , comme plus haut, la définition du symbole $g_X(a)$ au cas où a est algébrique sur K, en remplaçant K par une extension de degré fini convenable. Par linéarité, on définit ensuite $g_X(\mathfrak{a}) = (X, \mathfrak{a})$ pour tout cycle $\mathfrak{a}$ de dimension et de degré nuls, rationnel sur K . On prolonge enfin, par fonctorialité, comme au n. 5, la définition du symbole $(X, \mathfrak{a})$ au cas de diviseurs sur une variété V complète sans point multiple quelconque. Ce symbole est encore bilinéaire en X et $\mathfrak{a}$, et invariant par tout K-isomorphisme.

On généralise ainsi toutes les définitions introduites au n. 5. Toutefois, la méthode ci-dessus ne redonne pas, dans le cas des corps de fonctions, les propriétés de périodicité, ni le fait que les valeurs des symboles introduits sont rationnelles.

On peut aussi se demander s'il est encore possible, dans le cas général, d'interpréter le symbole $(X, \mathfrak{a})$ au moyen de la théorie des intersections et si on peut encore l'exprimer sous forme d'une somme de termes locaux.

Pour $v \in M$ ultramétrique, on peut introduire un modèle minimal local B de A relativement à v , et considérer le schéma $\mathcal{S}_v(B)$, sur l'anneau local de v, attaché à B . Utilisant la théorie des intersections sur $\mathcal{S}_v(B)$, on définit, en procédant comme au n. 6, les symboles locaux $(X, \mathfrak{a})_v$ et $j_v(X, \mathfrak{a})$.

On obtient encore $(X, \mathfrak{a}) = \sum_v (X, \mathfrak{a})_v$, mais à la condition d'introduire des termes complémentaires, respectivement associés aux valeurs absolues $v \in M$ archimédiennes. Dans ([6] , III, 7), il est indiqué comment ces termes peuvent s'interpréter en utilisant la théorie des fonctions thêta.

Il existe par ailleurs une manière directe de définir le symbole $(X, \mathfrak{a})_v$ ci-dessus valable à la fois pour les deux sortes de valeurs absolues (ultramétriques on archimédiennes) . On commence par poser, comme précédemment, $(X, \mathfrak{a})_v = v(X(\mathfrak{a}))$ lorsqu'on a $X \sim 0$. On montre ensuite qu'il existe une et une seule façon de prolonger cette définition au cas où X est quelconque, de sorte que le symbole dépende bilinéairement de X et de $\mathfrak{a}$ et que, de plus, les conditions suivantes soient satisfaites :

(i) - Pour tout K-morphisme $A \longrightarrow B$, et pour tout diviseur Y sur B , on a $(\varphi^{-1}(Y), \mathfrak{a})_v = (Y, \varphi(\mathfrak{a}))_v$.

(ii) - Désignant par $\mathfrak{U}$ l'ouvert de A complimentaire de X, et par a_o un point de $\mathfrak{U}_K$, la fonction. $a \rightarrow (X, (a)-(a_o))$ sur $\mathfrak{U}_K$ est localement bornée, i.e. est bornée sur tout sous-ensemble borné (pour la topologie définie par v) de $\mathfrak{U}_K$.

La définition peut encore être prolongée au cas des variétés quelconques . Le demonstration de l'unicite est elementaire. Celle de l'existence nécessite un passage à la limite faisant invervenir la notion de "quasi-fonction" voisine de celle de "distribution" au sens de Weil ([6], II, 9) .

BIBLIOGRAPHIE

[1] S. Lang : Diophantine geometry, Interscience Tracts, New York 1962

[2] ______ : Diophantine approximation on toruses , Amer. J. Math. 86 (1964) .

[3] I. Manin : Izvestya 28, 6 (1964) .

[4] A. Néron : Note aux C.R. Acad. Sc. Paris, t. 226 , p. 1781-1783 (1948).

[5] ________: Modéles minimaux des variétés abéliennes. sur les corps locaux et globaux, Publ. Inst. Hautes Et. Scientifiques. 21 (1964) .

[6] ________: Quasi-fonctions et hauteurs sur les variétés abéliennes. Ann. of Math. 82, 2(1965).

[7] ________ : Degré d'intersection en géometrie diophantienne, Proc. Du Congrès Internation. de Mathematique , Moscou (1966).

[8] A. Weil :: Foundations of algebraic geometry Amer. Mah. Soc. Colloquium Publ. , 29 (1962) .

[9] ________ : Arithmetic on algebraic variéties Ann. of Math. 78, 3 (1966) .

CENTRO INTERNAZIONALE MATEMATICO ESTIVO

(C.I.M.E.)

A. SEIDENBERG

REPORT ON ANALYTIC PRODUCTS

Corso tenuto a Varenna (Como) dal 7 al 17 Settembre 1969

Report on Analytic Products

by A. Seidenberg (University of California - Berkeley) (*)

1. For purposes of the present report, we shall feel free to consider situations as special as possible compatible with an exposition of the main issues.

We say that a complete local ring $\mathcal{O}$ is an analytic product if it is of the form $\mathcal{O}_1$ [[u]] with u a non-unit analytically independent over the subring $\mathcal{O}_1$.

Then $\mathcal{O}_1 \simeq \mathcal{O}/(u)$ is also complete, and since u is not a zero divisor, dim $\mathcal{O}_1$ = dim $\mathcal{O}$ -1. If, as will always be the case for us, there is a base field k present in $\mathcal{O}$, we also suppose it (in the definition) to be in $\mathcal{O}_1$.

Let V be an algebroid variety : V is given by an ideal $\mathfrak{a} \neq (1)$ in a ring of formal power series $k[[X_1,\ldots,X_n]]$ over a field k; let $k[[X_1,\ldots,X_n]]/\mathfrak{a} = k[[x_1,\ldots,x_n]]$

We say that V is an analytic product if $k[[x_1,\ldots,x_n]] = k[[x'_1,\ldots,x'_n]]$ with $(x_1,\ldots,x_n)$, $(x'_1,\ldots,x'_n)$ analytically related over k and x'_n is analytically independent over $h[[x'_1,\ldots,x'_{n-1}]]$.

If V is an analytic product, then obviously its ring $\mathcal{O} = h[[x]]$ is an analytic product ; the converse is also true, as one easily proves on the basis of the immediately following remarks.

Let V/k be an algebraic variety and W/k a subvariety.

We say that V is an analytic product at, or along, W if the completion of the local ring of V at W is an analytic product.

If $\mathcal{O} = \mathcal{O}_1$, [[u]] is an analytic product, then obviously there is an integral derivation of $\mathcal{O}$ over k , i.e., one sending $\mathcal{O}$

*Written in Rome , Italy, Summer 1969. Partially supported by the Centro Internazionale Matematico Estivo (C.I.M.E.).

into itself, with $Du = 1$ and $D\mathcal{O}_1 = 0$.

Moreover, if $a = a_o + a_1 u + \dots$ is the expansion of an $a \in \mathcal{O}$ according to $\mathcal{O} = \mathcal{O}_1 [[u]]$, then obviously $a \longmapsto a_o$ is an endomorphism of $\mathcal{O}$ onto $\mathcal{O}_1$, with (u) as kernel.

If k is of characteristic 0, as we assume throughout, a simple calculation shows that this endomorphism can be written as the operator $1 - uD + (u^2/2!) \cdot D^2 - \dots$ ($= e^{-uD}$).

Conversely, if D is an integral derivation over k of a complete local ring $\mathcal{O}$ with $Du = 1$ for a non-unit $u \in \mathcal{O}$, then one can recover the situation $\mathcal{O} = \mathcal{O}_1 [[u]]$, with $D\mathcal{O}_1 = 0$; see [2] or [6] .

Thus we can say :

<u>$\mathcal{O}$ is an analytic product if and only if its maximal ideal m is not differential</u> (i.e., $Dm \not\subset m$ for some D with $D\mathcal{O} \subset \mathcal{O}$).

Though it takes a little more work, a similar result holds for varieties : <u>if k [x] is the ring of an affine variety V/k, then the differential primes in k [x] correspond precisely to the subvarieties of V at, or along, which V is not an analytic product.</u>

This is one of the points of our paper [2] , where some of the apparatus for dealing with analytic products can be found.

Let m, $\mathfrak{a}$ be ideals in a polynomial ring k [X] with $m \supset \mathfrak{a}$, say the prime ideals of varieties U, V defined over k ; and let k' be an extension field of k ,.

One checks that $m/\mathfrak{a}$ is differential over k if and only if $k'[X]m/k'[X]\mathfrak{a}$ is differential over k'.

Thus this differential condition is a geometric condition in the sense of Weil's <u>Foundations.</u>

In terms of analytic products : <u>the condition that a variety</u>

V be an analytic product along a subvariety U is a geometric condition in the sense of Weil.

Precisely the same considerations hold for algebroid varieties U, V.

Therefore, in proving that an algebroid variety V is an analytic product along an algebroid subvariety U, we may arbitrarily extend the base field k.

In particular, we may assume that card k is greater than aleph null.

(Remark. In the case of an affine variety $k[x] = k[X]/\mathfrak{a}$ extending the base field k to k' yields the ring $k'[x] = k'[X]/\mathfrak{a}$, which is the tensor product $k' \otimes_k k[x]$. In the case of $k[[x]] = k[[X]]/\mathfrak{a}$, $k'[[x]] = k'[[X]]/\mathfrak{a}$ is not the tensor product $k' \otimes_k k[[x]]$, but rather a completion thereof.

However, if $\{\omega_i\}_{i \in I}$ is a linear basis of k'/k we can sometimes write infinite linear combinations

$$\sum_{i=1}^{\infty} \omega_i P_i(x), \quad P_i \in k[[x]], \quad \text{namely, if } P_i \longrightarrow 0 \text{ as}$$

$i \longrightarrow \infty$.

With this point in mind, the proofs of the above statement on U, V in the algebroid and algebraic cases are about the same.)

The assumption that card k is greater than aleph null will come in as follows. In algebroid geometry a problem will usually involve aleph null parameters, just as in algebraic geometry a problem usually involves a finite number of parameters.

These parameters are related through a set of equations.

In algebraic geometry, if a problem, i.e., a set of equations, has a solution in an extension field of an algebraically closed field k, then it already has a solution in k, by Hilbert's Nullstelleusatz.

By a theorem of Krull [1] , precisely the same can be said for a set of equations in infinitely many unknowns provided card k is greater than the cardinal number of the unknowns.

We come now to the main problem.

Let $F(X,Y,Z) = 0$ be an algebroid surface centered at the origin. In the case of interest, there will be a 1-dimensional singular component having a simple point at the origin.

By an analytic transformation over k we may suppose this component is the Z-axis. If $F = 0$ is an analytic product at the origin, then $F = 0$ is fibered in an obvious way near the origin : the section $Z = c$ centered at (o,o,c) is analytically equivalent to the section $Z = o$ at (o,o,o).

We need not enter into details, as we are concerned, rather, with the converse : Assuming that the sections by $Z = 0$ and $Z = c$, c indeterminate, are analytically equivalent (over a universal extension field Ω of $k((c))$, say), does it follow that $F = 0$ is analytically a product at the origin ?

The answer is yes (in characteristic 0 ; in characteristic $p \neq 0$, the surface $X^p + Z Y^p = 0$ provides a counter - example).

In the classical case that k = complex number field, we can consider sections $Z = c$ near $Z = 0$.

In the abstract case, we cannot do this, but can only allow $c = 0$ or c indeterminate.

In the latter case, we may think of $Z = c$ as being infinitely near $Z = 0$.

Our assumption can be stated as follows :

$F(X,Y,c) = F(\bar{X},\bar{Y},o)\,E$, where X,Y and $\bar{X},\bar{Y}$ are analytically related over Ω and E is a unit in $\Omega[[X,Y]]$.

A. Seidenberg

We are going to show that there are similar formulae over $k[[c]]$ reducing to the identity for $c = 0$.

Once this is done, it will be a mere matter of writing Z for C to get formulae showing that $F = 0$ is an analytic product.

Let $F(\Theta; X, Y)$ be the power-series in X, Y with indeterminate coefficients.

Then $F(X,Y,c)$ and $F(X,Y,o)$ can be written as $F(a_c;X,Y)$ and $F(a_o;X,Y)$, thereby displaying the coefficients; $a_o = a_c \mid c = 0$.

Let $E = \Lambda_o + \dots$ be another power series with indeterminate coefficients and let $\overline{X} = \Lambda_{11}X + \Lambda_{12}Y + \dots$, $\overline{Y} = \Lambda_{21}X + \Lambda_{22}Y + \dots$ with further indeterminates, which together with those of E we designate by Λ.

Write $F(\Theta(\Lambda); X, Y) = F(\Theta; \overline{X}, \overline{Y})\, E(\Lambda; X, Y)$.

This formula defines an action of Λ on Θ; and of any particular λ upon any particular θ.

Moreover, if λ is subjected to the requirement that E be a unit and that the substitution $(X,Y) \longrightarrow (\overline{X}(\lambda;X,Y), \overline{Y}(\lambda;X,Y))$ be analytic, i.e. to $\lambda_o(\lambda_{11}\lambda_{22} - \lambda_{12}\lambda_{31}) \neq 0$, then the action is a group action.

Let N be any positive integer. Let $F_N(\Theta;X,Y)$ be the sum of the terms in $F(\Theta; X,Y)$, of degree $\leq N$; and let E_N, $\overline{X}_N$, $\overline{Y}_N$ be similary defined.

Only a finite subset of the indeterminates Θ, Λ are involved, but we do not introduce further notation to indicate this.

Consider the congruence $\mathrm{mod}(X,Y)^{N+1}$ derived from the equalities in the last paragraph.

For example,

$$F_N(\Lambda(\Theta); X, Y) \equiv F_N(\Theta, \bar{X}_N, \bar{Y}_N)\ E_N(\Lambda; X, Y) \bmod (X, Y)^{N+1}$$

This, too, defines a group action, but this time that of an algebraic group, whose underlying set is an (abstract) affine space , upon an affine space.

Let $G = G_N$, $V = V_N$ designate this group and variety.

G operates regularly on V, i.e., the operation of G on V is an everywhere defined rational map of $G \times V$ into V.

Here G, V and the operation are defined over the rational number field Q.

Let $v \in V$ and g a $Q(v)$-generic point of G. $Q(v, g)$ is a regular over $Q(v)$, so $Q(v, g\, v)$ is also regular over $Q(v)$ and gv is a $Q(v)$-generic point of a variety $\vartheta(v)$ defined over $Q(v)$.

The variety $G \times \vartheta(v)$ is defined over $Q(v)$.

Let W be the subvariety of $G \times \vartheta(v)$ having $(g, g\, v)$ as $Q(v)$-generic point.

The (geometric) projection of W on $\vartheta(v)$ is $\vartheta(v)$. W consists of the points $(\bar{g}, \bar{g}\, v)$ with $\bar{g} \in G$.

Thus the set-theoretic projection of W on $\vartheta(v)$ is the orbit of v, and as the set-theoretic projection contains a non-empty $Q(v)$-open subset of the geometric projection, we see that $\vartheta(v)$ is the union of the orbit of v and a proper $k(v)$-closed subset.

For every N, a_c lies in the orbit of a_c; it is understood that one speaks only of the appropriate portions of a_c and a_o.

Now (N being fixed), let θ be a k-generic point of $\vartheta(a_o)$ and let $\lambda = (\lambda 1, \ldots, \lambda m)$ take a_o into θ.

Consider the point (θ, λ).

Let $r = \text{d.t. } k(0, \lambda)/k(\theta)$.

Applying a suitable non-singular linear transformation over $k(\theta)$, we may assume $k(\theta, \lambda_1, \ldots, \lambda_{r+1}) = k(\theta, \lambda_1, \ldots, \lambda m)$. We take the transformation over Q, so that specializations of the new λ_i over k correspond to specializations of the old λ_i over k; we will not introduce new notation for the new λ_i.

Corresponding to the defining equation for λ_{r+1} over $k(\theta, \lambda_1, \ldots, \lambda_r)$ we get an irreducible polynomial equation $F(\Theta, \Lambda_1, \ldots, \Lambda_r, \Lambda_{r+1}) = 0$; and corresponding to $\lambda_{r+j} \in k(\theta, \lambda_1, \ldots, \lambda_{r+1})$, $j > 1$, we get a polynomial equation $F_{r+j} = 0$ with $F_{r+j} = \Lambda_{r+j}\, d(\Theta, \Lambda_1, \ldots, \Lambda_r) -$

$P_{r+j}(\Theta, \Lambda_1, \ldots, \Lambda_{r+1})$.

Let $H(\Theta, \Lambda_1, \ldots, \Lambda_r)$ be $d(\Theta, \Lambda_1, \ldots, \Lambda_r)$ times the leading coefficient of F regarded as a polynomial in Λ_{r+1}.

Then by an elementary argument, the specializations $(\bar\theta, \bar\lambda)$ of (θ, λ) over k with exception of those for which $H(\bar\theta, \bar\lambda_1, \ldots, \bar\lambda_r) = 0$ are just the point $(\bar\theta, \bar\lambda)$ satisfying $F = 0$, $F_{r+j} = 0$ $(j = 2\ldots)$, the equations satisfied by θ over k, and the inequality $H \neq 0$.

By imposing another condition $H_1(\Theta, \Lambda_1, \ldots, \Lambda_r) \neq 0$, it will be true that if $(\bar\theta, \bar\lambda)$ is a specialization of $(\theta, \lambda)/k$, then $\bar\theta$ will be in the orbit of a_o and $\bar\lambda$ will take a_o into $\bar\theta$.

Imposing still another condition $H_2(\Theta, \Lambda_1, \ldots, \Lambda_r) \neq 0$, we can arrange to have

$$\frac{\partial F}{\partial \Lambda_{r+1}}\Bigg|_{\Theta = \bar{\theta},\ \Lambda = \bar{\lambda}} \neq 0.$$

We incorporate $H_1 H_2$ into H as a factor.

With these preparations, if $(\theta, \lambda) \longrightarrow (\bar{\theta}, \bar{\lambda})$ and $H(\bar{\theta}, \bar{\lambda}_1, \ldots, \bar{\lambda}_r) \neq 0$, we can solve $F((\Theta - \bar{\theta}) + \bar{\theta}, (\Lambda_1 - \bar{\lambda}_1) + \bar{\lambda}_1, \ldots) = 0$ for $\Lambda_{r+1} - \bar{\lambda}_{r+1}$ as a power series in $\Theta - \bar{\theta}, \Lambda_1 - \bar{\lambda}_1, \ldots, \Lambda_r - \bar{\lambda}_r$; and then from the other equations we get $\Lambda_{r+j} - \bar{\lambda}_{r+j}$ as a power series in $\Theta - \bar{\theta}, \Lambda_1 - \bar{\lambda}_1, \ldots, \Lambda_r - \bar{\lambda}_r$.

Let $\bar{\lambda}$ take a_c into $\bar{\theta}_c$; then the coordinates of $\bar{\theta}_c$ are in $k[[c]]$ and $\bar{\theta}_c \big|_{c=0} = \bar{\theta}$.

Now in the last paragraph repalce $\bar{\theta}$ by $\bar{\theta}_c$ and $\bar{\lambda}_1, \ldots, \bar{\lambda}_r$ by any $\bar{\lambda}_1 + c(\ldots), \ldots, \bar{\lambda}_r + c(\ldots)$, say by $\bar{\lambda}_1, \ldots, \bar{\lambda}_r$.

Note that $\bar{\theta}_c$ is a specialization of θ/k, as $\bar{\theta}_c$ is in the orbit of a_o.

We also have $H(\bar{\theta}_c, \bar{\lambda}_1, \ldots, \bar{\lambda}_r) \neq 0$, as this reduces to the previous inequality for $c = 0$.

Next we want to solve

$$F(\bar{\theta}_c, \bar{\lambda}_1, \ldots, \bar{\lambda}_r, (\Lambda_{r+1} - \bar{\lambda}_{r+1}) + \bar{\lambda}_{r+1}) = 0$$

for $\Lambda_{r+1} - \bar{\lambda}_{r+1}$ and can do this since

$$\frac{\partial F}{\partial \Lambda_{r+1}}\Bigg|_{c = 0,\ \Lambda_{r+1} = \bar{\lambda}_{r+1}} \neq 0$$

getting a power series in $ck[[c]]$; let $\lambda_{r+1,c} = \bar{\lambda}_{r+1} + c(\cdot\cdot)$.

Then we can solve the remaining

$F_{r+j}(\bar{\theta}_c, \bar{\lambda}_1, \ldots, \bar{\lambda}_{r+1}, \Lambda_{r+j}) = 0$, getting

$\lambda_{r+j,c} = \bar{\lambda}_{r+j} + c(\cdot\cdot)$.

Now we have a specialization $(\bar{\theta}_c, \lambda_c)$ of (θ, λ) over k.

The transformation λ_c followed by the inverse of $\bar{\lambda}$ takes a_o into a_c.

Since $\lambda_c \big|_{c=0} = \bar{\lambda}$, this last transformation is of the form $\bar{X} = (1 + c(\cdot\cdot)) X + c(\cdot\cdot) Y + (\cdot\cdot)X^2 + \ldots$,

$\bar{Y} = c(\cdot\cdot) X + (1 + c(\cdot\cdot)) Y + (\cdot\cdot) X^2 + \ldots$, $E = 1 + (\cdot\cdot) X + (\cdot\cdot)Y + \ldots$,

with the $(\cdot\cdot)$ in $k[[c]]$.

<u>Thus up to terms of degree N +1 in X and Y we have a transformation of the desired kind.</u>

The idea of this last part of the proof was this : by bringing a_o and a_c into sufficiently general positions $\bar{\theta}$, $\bar{\theta}_c$ of their orbit, we can compare what goes on at $\bar{\theta}$ and $\bar{\theta}_c$, whereas we cannot do this directly for a_o and a_c because of their possibly special positions.

So far we have been working with a given positive N.

For each Λ_i let us write $\Lambda_i^{(o)} + \Lambda_i^{(1)}c + \ldots$

Consider the equations $F(a_c, \bar{X}, \bar{Y}) = F(a_o; \bar{X}, \bar{Y})\, E(\Lambda; X, Y)$,

$\bar{X} = (1+c(\cdot\cdot))X + c(\cdot\cdot)Y + \ldots$, $\bar{Y} = c(\cdot\cdot)X + (1 + c(\cdot\cdot))\, Y + \ldots$,

$E = (1 + c(\cdot\cdot)) + (\cdot\cdot)\, X + (\cdot\cdot)\, Y + \ldots$ (i.e., we impose the conditions $\Lambda_o^{(o)} = 1$, $\Lambda_{11}^{(o)} = \Lambda_{22}^{(o)} = 1$, $\Lambda_{12}^{(o)} = \Lambda_{21}^{(o)} = 0$).

This is equivalent to an infinite set of equations over k in the $\Lambda_i^{(j)}$; and we seek a solution in k.

Modulo $(X, Y)^{N+1}$ we get a system of equations having a solution in k.

Write the equations in the $\Lambda_i^{(j)}$ in the form $P_k(\Lambda) = 0$.

Any finite number of these appear when we go modulo $(X, Y)^{N+1}$: so the ideal generated by them in the appropriate polynomial ring over k does not = (1).

Hence also the ideal generated by all the $P_k(\Lambda)$ in $k[\Lambda]$ does not = (1).

Hence the $P_k(\Lambda) = 0$ have a solution in an extension field of k.

Applying the generalized Hilbert Nullstellensatz and our previous remarks on card k , we complete the proof.

2. The result of § 1 can be generalized in two directions.

In the first, an arbitrary ideal $\mathfrak{a} \neq (1)$ takes over the role of (F) of § 1.

The hyperplane $X_1 = 0$ will take the role of $Z = 0$.

The role of the Z-axis is taken by a prime ideal n containing $\mathfrak{a}$, or rather by its associated algebroid variety V(n).

By the generic point $P : (c_1, \ldots c_n)$ of V(n) we mean the point having the residues of $X_1, \ldots, X_n$ in $k[[X]]/n$ as coordinates.

We assume that the r-dimensional algebroid variety V(n) has a simple point at the origin and that $X_1 = 0$ is not tangent there to V(n).

By the first assumption we mean that $k\ [[X]]\ /\mathfrak{n} = k\ [[c]]$ is an r-dimensional regular local ring; and by the second, that $X_1/\mathfrak{n} = c_1$ is not in the square of the maximal ideal of $k\ [[X]]\ /\mathfrak{n}$ (hence $r > 0$).

We compare the algebroid variety centered at the generic point $(c_1, \ldots, c_n)$ of $V(\mathfrak{n})$ and cut out on $V(\mathfrak{a})$ by $X_1 - c = 0$. with the algebroid variety centered at the origin and cut out on $V(\mathfrak{a})$ by $X_1 = 0$. Then we have:

Theorem A. Let k be algebraically closed, of characteristic 0, $\mathfrak{a}$ an ideal in the ring of formal power series $h\ [[X_1, \ldots, X_n]]$, $\mathfrak{a} \neq (1)$, and $\mathfrak{n}$ a prime ideal containing $\mathfrak{a}$ such that $V(\mathfrak{n})$ has a simple point at the origin and is not tangent there to $X_1 = 0$.

Assume that the algebroid variety centered at the generic point $(c_1, \ldots, c_n)$ of $V(\mathfrak{n})$ and cut out on $V(\mathfrak{a})$ by $X_1 - c_1 = 0$ is analytically equivalent over an extension field Ω of $h((c))$ to the algebroid variety centered at the origin and cut out on $V(\mathfrak{a})$ by $X_1 = 0$, or, in ring-theoretic terms, that

$\Omega\,[[X - c]]\ /\ \Omega\,[[X - c]] \cdot (\mathfrak{a}, X_1 - c_1)$ and

$\Omega\,[[X]]\ /\ \Omega\,[[X]]\ (\mathfrak{a}, X_1)$ are isomorphic over Ω.

Then $V(\mathfrak{a})$ is an analytic product at the origin.

In § 1, $\mathfrak{n}$ is 1-dimensional, and in the course of the proof of Th. A we establish the following ;

Lemma. Let $\mathfrak{a}$ and $\mathfrak{n}$ be as in the theorem.
Then there exists a 1-dimensional prime ideal $\mathfrak{n}_1$ having

the properties stated in the theorem.

With the lemma in mind, the proof of Th. A is, in broad outline, the same as the proof of the special case in § 1 (one may assume that $V(n_1)$ is the X_1-axis, one expresses the analytic equivalence of the sections in terms of a group action, etc.)

For details, see [4] .

Instead of a fibration of $V(\mathfrak{a})$ by hyperplane sections, we may more generally consider a fibration also by lower dimensional sections.

To introduce some terminology, what before we called an analytic product we will now call an analytic product with a line ; if $k[[X]]/\mathfrak{a} = k[[x]]$ is of the form $\mathcal{O}_1[[u, v]]$ with u, v analytically independent over a ring $\mathcal{O}_1$, containing k, then $V(\mathfrak{a})$ will be said to be an analytic product with a plane, etc.

Theorem B. Let the field k and the ideal $\mathfrak{a}$ in $k[[X_1, \ldots, X_n]]$ be as in Theorem A, and let n be an r-dimensional prime ideal containing $\mathfrak{a}$ such that $V(n)$ has a simple point at the origin and is not tangent there to the linear space

$L^{n-r}: X_1 = 0, \ldots, X_r = 0,$

Assume that the algebroid variety centered at the generic point $(c_1, \ldots, c_n)$ of $V(n)$ and cut out on $V(\mathfrak{a})$ by

$L_c^{n-r}: X_1 - c_1 = 0, \ldots X_r - c_r = 0$ is analytically equivalent over an extension field Ω of $k((c))$ to the algebroid variety cut out by L^{n-r}.

Then $V(\mathfrak{a})$ is analytically a product with an L^r at the origin. (See [4] .)

3. Perhaps we should still say something about our paper [2] , which was the starting point of our considerations.

For the present, we consider only the case of the ring $k [x]$ of an irreducible variety V/k.

The problem posed in [2] was to locate the differential primes of $k [x]$.

In Th. 5 we showed that the minimal differential primes (not counting (0)) correspond precisely to the component of the singular locus of V.

As previous by remarked, the non-differential primes p correspond precisely to the subvarieties W along which V is an analytic product.

In this way we come to study algebraico - algebroid problems.

Most, or even nearly all, of these problems have a purely algebroid counterpart .

For example, Th. 5 can be stated, and is true, for algebroid varieties V, too.

In fact, all of the results of [2] remain correct for algebroid varieties.

Moreover, with one exception, the proofs are simple modifications of the proofs in the algebraic case.

The one exception is Th. 4, which we will not quote, but the difficulty is this : In Th. 4, we used the well-known technique of "reduction to dimension zero" (see [2] , p. 27, bottom).

A. Seidenberg

In the case of an affine domain $k[x_1, \ldots, x_n]$ one adjoins some of the x_i to the base field k; the result is an affine ring of dimension zero over a larger field.

This idea does not work for a complete local domain $k[[x_1, \ldots, x_n]]$, and some augmentation of technique is necessary.

In [2], several questions were mentioned. Of these, Question 1 refers to the problem dealt with in [4] and above in §§ 1, 2.

Question 2 has been answered in [5].

Some preliminary thoughts on Question 3 led to our paper [3].

But the main question posed remains open: we know where the minimal differential primes are and can say something about the k-rational points which determined differential primes (see Th. 13 of [2]), but for example, we have no adequate, or even <u>any</u> notion of where the 1-dimensional primes of $k[x]$ are.

REFERENCES

[1]. W. Krull, "Jacobsonsche Ringe, Hilbertscher Nullstellensatz, Dimensionstheorie", Math. Z., vol. 54 (1961), pp. 354-387.

[2]. A. Seidenberg, "Differential Ideals in Rings of Finitely Generated Type", Amer. J. Math., vol. 89 (1967), pp. 22-42.

[3]. ________, "Reduction of Singularities of the Differential Equation Ady = Bdx", Amer. J. Math., vol. 90 (1968), pp. 248-269.

[4]. ________, "Analytic Products". Scheduled to appear in the April 1969 issue of the Amer. J. Math.

[5]. ________, "On Analytically Equivalent Ideals". To appear in the Institut des Hautes Etudes Scientifiques, Zariski volume.

[6]. O. Zariski, "Studies in Equisingularity, I", Amer J. Math., vol. 87 (1965), pp. 507-536.

CENTRO INTERNAZIONALE MATEMATICO ESTIVO

(C. I. M. E.)

C.S. SESHADRI

MODULI OF π- VECTOR BUNDLES OVER AN ALGEBRAIC CURVE

Corso tenuto a Varenna (Como) dal 7 al 17 Settembre 1969

MODULI OF π—VECTOR BUNDLES OVER AN ALGEBRAIC CURVE

By C. S. Seshadri (Tata Institute, Bombay)

Introduction : Let X be a smooth algebraic curve, proper over $\mathcal{C}$ the field complex numbers (or equivalently a compact Riemann surface) of genus g. Let J be the Jacobian of X ; it is a group variety of dimension g and its underlying set of points is the set of divisor classes (or equivalently isomorphic classes of line bundles) of degree zero.

It is a classsical result that the underlying topological space of J can be identified with the set of (unitary) characters of the fundamental group $\pi_1(X)$ into $\mathcal{C}$ (i.e. homomorphisms of $\pi_1(X)$ into complex numbers of modulus one) and therefore $J = S^1 \times \ldots \times S^1$, g times, as a topological manifold S^1 being the unit circle in the complex plane.

The purpose of these lectures is to show how this result can be extended to the case of unitary representations of arbitrary rank of Fuchsian groups with compact quotient.

Given a representation $\rho : \pi_1(X) \longrightarrow GL(n, \mathcal{C})$, one can associate to ρ in a natural manner (as will be done formally later) a vector bundle (algebraic or holomorphic) on X; let us call such a vector bundle unitary if ρ is a unitary representation.

It is easy to show that two vector bundles V_1, V_2 associated to unitary representations ρ_1, ρ_2 are isomorphic (in the algebraic or holomorphic sense) if and only if ρ_1 and ρ_2 are equivalent as representations.

We will prove the following results : Suppose that the genus of X is $\geqslant 2$.

Then we have :

(1) The unitary bundles on X can be characterized algebraically.

To be more precise, let us call a vector bundle V on X of degree zero on X stable (resp. semi-stable) (this definition is due to Mumford) if for every proper sub-bundle W of V, one has deg W < 0 (resp. deg W ≤ 0).

Then one has the following : a vector bundle V on X is unitary if and only if it is a direct sum of stable bundles (of degree 0) and V is stable if and only if the corresponding unitary representation is irreducible unitary.

(2) On the equivalence classes of unitary representations of a given rank of $\pi_1(X)$ (equivalently, on the isomorphic classes of unitary vector bundles on X of a given rank), there is a natural structure of a normal projective variety.

It can be shown that there are stable bundles on X of arbitrary rank if $g \geq 2$ and that in any case, they form a Zariski open subset of any algebraic family of vector bundles.

Let $\tilde{X}$ denote a simply connected covering of X.

Then $\pi = \pi_1(X)$ can be identified with a proper discontinuous group of automorphisms acting freely on $\tilde{X}$.

It is easily seen, since π operates freely on $\tilde{X}$, that the study of (holomorphic) vector bundles on X is equivalent to the study of π-vector bundles on $\tilde{X}$, a π-vector bundle on $\tilde{X}$ being a (holomorphic) vector bundle E on $\tilde{X}$ together with an action of π on E compatible with its actio on X.

From this point of view the results (1) and (2) admit of generalisations and in fact in the following, we shall be concerned with the following more general situation :

Let $\tilde{X}$ be a aimply connected Riemann surface and Γ a proper discontinuous group of automorphisms of $\tilde{X}$ such that $Y = \tilde{X} \bmod \Gamma$ is compact (the action of Γ is not supposed to be free). It is well-known that Γ has a normal subgroup of finite index Γ_o in Γ such that Γ_o operates freely on $\tilde{X}$.

Let $X = \tilde{X} \bmod \Gamma_o$ and $\pi = \Gamma \bmod \Gamma_o$. Then there is a canonical action of π on X such that $Y = X \bmod \pi$.

It is easily seen that the study of -vector bundles on X is equivalent to the study of π-vector bundles on $\tilde{X}$ and thus the study of Γ-vector bundles on $\tilde{X}$ can be said to be an algebraic problem.

Given a representation ρ of Γ into $GL(r, C)$, there is a natural Γ-vector bundle on $\tilde{X}$ (of rank r) and consequently a π-vector bundle E on X associated to ρ.

Let us call a π-vector bundle E on X π-unitary (resp. irreducible π-unitary) if it is π-isomorphic (i.e. isomorphic in the category of π-vector bundles) to a π-vector bundle associated to a unitary (resp. irreducible) representation of Γ.

As in the case of a free action, if E_1, E_2 are two π-unitary vector bundles on X associated to unitary representations ρ_1, ρ_2 of Γ, then E_1 is π-isomorphic to E_2 if and only if the representations ρ_1, ρ_2 of Γ are equivalent.

Let us call a π-vector bundle E on X of degree zero, π-semi-stable if the underlying vector bundle is semi-stable (as defined above) and π-stable if for every proper π-sub-bundle W of E, we have $\deg W < 0$.

Let us call two π-vector bundles E_1 and E_2 on X locally isomorphic at $x \in X$ if there is a neighbourhood U of x invariant under π_x-the isotopy group of π at x, such that the restrictions of

E_1 and E_2 to U are π_x-isomorphic; let us call E_1 and E_2 <u>locally isomorphic</u> if they are locally isomorphic at every point of X (unlike the usual case, any two π-vector bundles of the same rank need not be locally isomorphic).

We can thus speak of the <u>local type</u> of a π-vector bundle on X.

With these definitions, these lectures are devoted to the proof of the following results generalising (1) and (2) above:

Suppose that genus of Y = $\tilde{X}$ mod Γ is ≥ 2. Then

<u>Theorem I.</u> A π-vector bundle on X of degree zero is π-stable if and only if it is irreducible π-unitary (cf. Th. 4, Chap. II).

<u>Theorem II.</u> On the space of isomorphic classes of -unitary vector bundles on X of a fixed local type τ, there is a natural structure of a normal projective variety (cf. Th. 5, Chap. II).

We shall now give a brief outline of the proofs of the above theorems.

One proves directly that a π-unitary (resp. irreducible π-unitary) vector bundle E on X is π-semi-stable (resp. π-stable) of degree zero; further given an analytic family $\{E_t\}_{t \in T}$ of π-semi-stable vector bundles on X parametrized by an analytic space T, the subset T_o of points $t \in T$ such that E_t is irreducible π-unitary is <u>closed</u> in T. (cf. Prop. 10, Chap. II).

The one shows that given an analytic family of π-vector bundles $\{E_t\}_{t \in T}$ on X the subset T_o of T such that E_t is irreducible π-unitary is <u>open</u> in T (cf. prop. 7, Chap. II).

For proving this one proceeds as follows.

Let ρ be an irreducible unitary representation of Γ of rank r and E the associated π-bundle on X.

Then it is proved that if U is the real analytic space of all unitary representations of Γ of rank r, then U is smooth (i.e. a manifold) in a neighbourhood of ρ and the dimension of the topological manifold constituted by this smooth neighbourhood is

$$2 \dim_{\mathscr{C}} H^1(X, \pi, E^* \otimes E) + \dim K_r - 1$$

where K_r-denotes the group of unitary matrices of rank r, E the dual vector bundle of E and $H^1(X, \pi, E^* \otimes E)$ denotes a naturally defined first cohomology group of $E^* \otimes E$ in the category of abelian π-sheaves on X. (cf. Th. 3 and Cor. 2, Th. 3, Chap. 1).

Since K_r acts by inner conjugation on U and we have a natural induced free action of $PK_r = K_r$ modulo its centre consisting of scalar matrices on U, we have a manifold V such that $\dim V = 2 \dim_{\mathscr{C}} H^1(X, \pi, E^* \otimes E)$ and which represents a nice local module of unitary representations around the point ρ (cf. Cor. 1, Th. 3, Chap. I).

On the other hand it is known after the work of Kodaira-Spencer that there is a nice local module D of vector bundles around the point E and which is a complex analytic manifold of complex dimension $\dim_{\mathscr{C}} H^1(X, \pi, E^* \otimes E)$; we obtain this not by using the work of Kodaira-Spencer but as a consequence of studying a suitable Quot scheme in the sense of Grothendieck and which is in any way required for the proof of Th. II (cf. Cor., Prop. 6, Chap. II).

Thus we have

$$\dim_{\mathscr{C}} D = 2 \dim_{\mathscr{R}} V$$

and this implies easily that T_o is open in T as required above.

Thus to prove Theorem I, it suffices to produce a connected tot all family for all π -semi-stable vector bundles of a fixed local type.

This fact is a consequence of the considerations used to prove Theorem II.

To prove Theorem II, one constructs an algebraic family of π -vector bundles on X parametrized by a smooth, irreducible, quasi-projective variety R^{τ} containing all π -semi-stable vector bundles of a fixed local type τ and an action of a reductive algebraic group H on R^{τ} such that the orbits in R^{τ} under H correspond precisely to isomorphic classes of π -vector bundles (cf. Prop. 6, Chap. II).

This is done by using the Quot schemes of Grothendieck.

The real difficulty now starts since as Mumford and Nagata have shown, the orbit spaces under the action of an algebraic group need not exist in the category of algebraic schemes even in good cases.

Now one follows the ideas of Mumford (cf. [8]) by reducing this question to a problem of constructing orbit spaces for a product of Grassmannians under the diagonal action of the projective group and connects the stable (resp. semi-stable) vector bundles with the stable (resp. semi-stable) points of this product of Grassmannians under the action of the projective group (cf. Prop. 9 and Cor. 2, Prop. 9 as well as § 3, Chap. II).

Then Theorem II follows from these considerations.

The Theorems I and II above, in the particular case of a free action of Γ are the main results of [12] and [17] respectively.

The proofs outlined above for the general case are substan=

tially the same as in [12] and [17] .

Some technical improvements upon the proofs of [12] and [17] are included here.

The fact that the morphism χ of Cor. 2, Prop. 9, Chap. II, is proper is taken from [18] and this makes the proof of Theorem II more direct than in [17] and avoids the use of § 4 and § 5 of [17] .

Another fact worth mentioning is the proof (due to S. Ramanan) given here that an irreducible unitary bundle is stable.

This is more direct than that of [12] .

The problem of constructing moduli of π -vector bundles over an algebraic curve, was raised for the first time by A. Weil in [19] .

In fact, most of the material in § 2, Chap. I is to be found in [19] ; in the presentation of this material i.e. § 1 and § 2, Chap. I, we have followed the exposition of Grothendieck [4] of this paper of Weil.

The existence of a quasi-jective moduli space for stable bundles was first proved by Mumford [8] .

C.S. Seshadri

Chapter I

Unitary π -bundles

1. Generalities on π -bundles :

Let X be a complex analytic space and π a discontinuous group of automorphisms of X i.e. π acts as a group of analytic automorphisms of X and satisfies

(i) $\forall\, x \in X$, the isotropy group π_x at x is finite $\exists$ an open neighbourhood U_x of x such that $\pi_x U_x = U_x$ and $U_x \cap gU_x = \emptyset$ for $g \notin \pi_x$.

Then it can be shown that $Y = X/\pi$ ($p: X \longrightarrow Y$ canonical map) has a natural structure of a complex analytic space (cf. [2]).

(In fact, we require this only for the case when X is a manifold of dimension one and in this case it is easy to see that Y is a manifold of dimension one and the image of the canonical map $\pi_x \longrightarrow \mathrm{Aut}\, U_x$ is cyclic).

A fibre space $p : E \longrightarrow X$ (or an X-analytic space) is called a π -fibre space (or a π -analytic space) over X if π operates on E and $p : E \longrightarrow X$ is a π -morphism i.e. commutes with the operations of π .

We say that a sheaf $\mathcal{G}$ (of sets, groups, rings etc.) on X is a π -sheaf if the corresponding étale space over X associated to $\mathcal{G}$ is a π-fibre-space over X (note that the étale space acquires canonically a structure of an X-analytic space).

The definition of a π - sheaf can also be done by a procedure resembling a prescheaf datum; we leave the details.

Now π -sheaves on X (resp. of sets, groups, rings etc.) form a category under morphisms which commute with the action of π .

Given a Y-analytic space, $E \longrightarrow Y$, the base change of $E \longrightarrow Y$ by $p : X \longrightarrow Y$, namely $E \times_Y X$ is a π-analytic space over X, induced by the canonical operation of π on $E \times_Y X \subset E \times X$, taking the trivial action of π on E.

We denote $E \times_Y X$ by $p^*(E)$.

If $E \longrightarrow Y$ is a local isomorphism (i.e. E is étale over Y), then $E \times_Y X$ is étale over X.

From this one concludes easily that if $\mathcal{G}$ is a sheaf (resp. of sets, groups, rings, etc.), then $p^*(\mathcal{G})$ has a natural structure of a π-sheaf over X (resp. of sets, groups, rings etc.).

Then $\mathcal{G} \longrightarrow p^*(\mathcal{G})$ defines a functor from the category of sheaves on Y (resp. of sets, groups, rings, etc.) into the category of π-sheaves on X (resp. of sets, groups, rings, etc.)

We note that we have a canonical functorial map $H^o(Y, \mathcal{G}) \longrightarrow H^o(X, p^*(\mathcal{G}))$ (i.e. π-invariant sections of $p^*(\mathcal{G})$ over X).

For every π-sheaf $\mathcal{G}$ on X, we denote by $p(\mathcal{G})$, the direct image of $\mathcal{G}$ by p (sections of $p_*(\mathcal{G})$ on an open $V \subset Y$ are sec= tions of $\mathcal{G}$ over $p^*(V)$).

We note that $p_*(\mathcal{G})$ acquires a natural structure of a π-sheaf on Y (taking the trivial action of π on Y).

We denote by $p_*^{\pi}(\mathcal{G})$ (the invariant direct image of $\mathcal{G}$) the subscheaf π-invariants of $p_*(\mathcal{G})$ ($p_*^{\pi}(\mathcal{G})$ is defined by the following prescheaf: to every open V in Y assign the π-invariant sections of $\mathcal{G}$ over $p^*(V)$).

We note that $p_*^{\pi}(G)$ is a sheaf of groups, rings etc. accor= ding as $\mathcal{G}$ is so.

Now $\mathcal{G} \longmapsto p_*^{\pi}(\mathcal{G})$ defines a functor from the category of

π-sheaves on X (resp. of sets, groups, rings) to the category of sheaves (resp. of sets, groups, rings) on Y.

If we denote by $H^o(X, \pi, \mathcal{G})$ the set of π-invariant sections of $\mathcal{G}$ over X, we have $H^o(X, \pi, \mathcal{G}) = H^o(Y, p_*^{\pi}(\mathcal{G}))$.

If E is a sheaf on Y, then $p_*^{\pi}(p^*(E))$ identifies canonically with E.

On the other hand, if $\mathcal{G}$ is a π-sheaf on X, then $p^*(p_*^{\pi}(\mathcal{G}))$ identifies with the subsheaf $\mathcal{G}^{\pi}$ of $\mathcal{G}$, whose stalk at x is $G_x^{\pi_x}$ -the subset of π_x-invariants of $\mathcal{G}_x$ under π_x.

Since $\mathcal{G} \longmapsto p_X^{\pi}(\mathcal{G})$ and $E \longmapsto p^*(E)$ are functors, we conclude easily that when π operates freely (so that $\pi_x = \text{Id.}$ for every $x \in X$), the functor p_*^{π} establishes an equivalence of the category of π-sheaves (resp. of sets, groups, rings etc.) on X with the category of sheaves on Y (resp. of sets, groups, rings etc.) (similarly for p^*).

We note that $\mathcal{O}_Y = p_*^{\pi}(\mathcal{O}_X)$, where $\mathcal{O}_X$, $\mathcal{O}_Y$ denote the structure sheaves of rings of X and Y respectively.

Similarly if G is a complex Lie group, then the sheaf $\mathcal{O}_X(G)$ of germs of analytic morphisms of X into G is a π-sheaf (of groups) and we have $p_*^{\pi}(\mathcal{O}_X(G)) = \mathcal{O}_Y(G)$.

Let $P \longrightarrow X$ be a principal fibre space (analytic) with structure a complex Lie group G.

We say that P is a π-principal fibre space with structure group G (or briefly a π-G bundle) if we are given an operation of π on P which commutes with that of G and induces the given operation of π on X.

We define in the obvious manner an isomorphism of two

(π -G) bundles and denote the isomorphism classes by $H^1(X, \pi\, \mathcal{O}_X(G))$.

We define similary the notion of an associated π -bundle to a principal (π -G) bundle on X.

We can thus speak of a π -vector bundle, namely the π -vector bundle associated to a π -GL(n) bundle.

We have a natural notion of a π -<u>homomorphism</u> between two -vector bundles and when we speak of the category of π -vector bundles, we take for morphisms π -homomorphisms.

Given a homomorphism $\rho : \pi \longrightarrow G$ of π into a complex Lie group G, we obtain in a natural manner à π -G bundle $P \longrightarrow X$ as follows : we take $P = X \times G$ and define the operation of π on P as $\alpha \circ (x, g) = (\alpha \circ x, \rho(\alpha)\, g), \alpha \in \pi$.

In particular if $\mathcal{G}$ = GL(V) (group of linear automorphisms of a finite dimensional vector space V over $\mathcal{C}$), we get a π -GL(V) bundle and its associated vector bundle will be referred to as the <u>π -vector bundle associated to the representation</u> ρ

The direct sum (as well as tensor product) of two π-vector bundle has a natural structure of a π -vector bundle.

The dual E^* of a π -vector bundle has a natural structure of a π -vector bundle.

If E_1, E_2 are two π -vector bundles associated to representations ρ_1, ρ_2 of π , then the π -vector bundle $E_1 \otimes E_1$ is associated to the representation $\rho_1 \otimes \rho_2$.

(Similarly we get a statement for the dual representation).

Let V_1, V_2 be the representation spaces of ρ_1, ρ_2 respectively.

Then we have a canonical homomorphism

$$\mathrm{Hom}_{\pi}(V_1, V_2) \longrightarrow \mathrm{Hom}_{\pi}(E_1, E_2)$$

where $\mathrm{Hom}_{\pi}(V_1, V_2)$ denotes the $\mathcal{C}$ linear space of π-homomorphisms of V_1, V_2 and $\mathrm{Hom}_{\pi}(E_1, E_2)$ denotes the $\mathcal{C}$-linear space of π-homomorphisms (analytic) of E_1 into E_2.

Proposition : Let us suppose that X is a connected complex manifold and that Y is compact.

Let ρ_1, ρ_2 be two unitary representations of π on (finite dimensional) vector spaces V_1, V_2 (i.e. ρ_1, ρ_2 leave invariant positive definite Hermitian forms on V_1, V_2 respectively).

Then the canonical homomorphism

$$\mathrm{Hom}_{\pi}(V_1, V_2) \longrightarrow \mathrm{Hom}_{\pi}(E_1, E_2)$$

is an isomorphism.

In particular, if ρ is a unitary representation on V and E the associated π-bundle, then the natural map

$$V^{\pi} \longrightarrow H^{o}(X, \pi, E)$$

is an isomorphism.

Proof. It suffices to prove the second assertion bacause of the following $\mathrm{Hom}_{\pi}(V_1, V_2) = \mathrm{Hom}_{\pi}(\mathcal{C}, V_1^* \otimes V_2)$ (π operating trivially on $\mathcal{C}$) $= (V_1^* \otimes V_2))^{\pi}$

Similarly $\mathrm{Hom}_{\pi}(E_1, E_2)$ is $H^o(X, \pi, E_1^* \otimes E_2)$.

Thus it suffices to prove that the canonical map $(V_1^* \otimes V_2)$ $H^o(X, \pi, E_1^* \otimes E_2)$ is an isomorphism.

Now E_1^*, $E_1^* \otimes E_2$ are unitary bundles and thus we are reduced to proving the last assertion.

Now $E = X \times V$ and thus a π-section s of E can be identified with a holomorphic map $F : X \longrightarrow V$

If $\| \ \|$ denotes a Hermitian metric on V invariant under

ρ, we see that the function $h : X \longrightarrow \mathcal{R}$, $h(x) = \|F(x)\|^2$ is π-invariant.

This implies that h goes down to a function $g : Y \longrightarrow \mathcal{R}$ and g is obviously continuous.

In particular g attains both its maximum and minimum at some points of Y.

This shown that h also attains its maximum and minimum at some points of X.

If we introduce a basis $\{e_i\}$ in V such that if $v = \sum z_i v_i, z_i \in C, \|v\|^2 = \sum |v_i|^2$, we have $F(x) = \sum F_i(x) e_i$ and $h(x) = \|F(x)\|^2 = \sum |F_i(x)|^2$ i.e. h(x) is a sum of squares of moduli of holomorphic functions on X.

It follows that h(x) is plurisubharmonic and since it attains its maximum at an interior point of the connected manifold X, h reduces to a constant map.

From this one concludes that the holomorphic map $F : X \longrightarrow V$ is a constant map bacause of the following :

Lemma 1 : Let X be a connected complex manifold and $h(x) = \sum_{i=1}^{r} |F_i(x)|^2$, where F_i is a holomorphic function on X.

Then if h is constant, the F_i also reduce to constant functions.

Proof. It is easy and left as an axercise.

Since $F : X \longrightarrow V$ is a constant map, we see immediately that $F(X) \in V^{\pi}$.

From this it is immediate that the map $V^{\pi} \longrightarrow H^0(X, \pi, E)$ is an isomorphism and the proposition is proved.

Corollary: Let E_1, E_2 be two π-vector bundles on X asso=

ciated to unitaty representation ρ_1, ρ_2 of π (X as in the proposi= tion).

Then E_1 is isomorphic to E_2 as π-bundles if and only the representations ρ_1 and ρ_2 are rquivalent.

Proof. This is an immediate consequence of the proposition.

Let P be a (π - G) bundle (G complex Lie group) on X such that the underlying principal G-bundle is trivial i.e. $P \cong X \times G$.

We can write the operation of π on $X \times G$ as follows (ta= king the operation of G on P to be on the right):

$$(*) \qquad \alpha \cdot (x, g) = (\alpha \cdot x, f_\alpha(x)\, g)$$

where $f_\alpha : X \longrightarrow G$ is a holomorphic map.

Writing down the conditions that π operates on $X \times G$, we obtain

$$f_{\alpha\beta}(x) = f_\alpha(\beta x)\, f_\beta(x), \quad \alpha, \beta \in \pi$$

i.e. $\alpha \longmapsto f_\alpha$ defines a 1-cocycle of π with values in $\Gamma(X, \mathcal{O}_X(G))$, for the canonical operation of π on $\Gamma(X, \mathcal{O}_X(G))$ (note that $f(x) \longmapsto f(\beta x)$, $\beta \in \pi$ defines an operation of π on $\Gamma(X, \mathcal{O}_X(G))$ on the right).

Conversely, given a 1-cocycle of π with values in $\Gamma(X, \mathcal{O}_X(G))$, we get an operation of π on $P = X \times G$ by ($*$), commuting with the action of G and thus P acquires a π-bundle structure.

Given two such (π -G) bundles P_1 and P_2 given by 1-cocy= cles $\{f_\alpha(x)\}$ and $\{g_\alpha(x)\}$, P_1 and P_2 are π-isomorphic if and only the cocycles are cohomologous i.e. $\exists$ a holomorphic map $F : X \longrightarrow G$ such that $F(\alpha x) = g_\alpha(x)\, F(x)\, f_\alpha(x)^{-1}$ do that the set of isomorphic classes of such bundles can be identified with the set $H^1(\pi, \Gamma(X, \mathcal{O}_X(G))$ (this has no structure of a group if G is non-abelian).

We see that π-bundles associated to representations of π are particular cases of these π-bundles.

Given any (π-G) bundle, we can choose a neighbourhood U_x for every $x \in X$ invariant under π_x such that the restriction of the underlying G-bundle to U_x is trivial and thus locally every (π-G) bundle is of this type.

Suppose now F is a coherent π-sheaf on X, then $p_*^{\pi}(F)$ is a coherent sheaf on Y.

(Choosing a neighbhourhood U_x of x invariant under π_x, it suffices to show that $p_*^{\pi_x}(F / U_x)$ is coherent on U_x / π_x and thus to show coherence of $p_*^{\pi}(F)$, we are reduced to the case of a finite group.

Further this fact is immediate when X is a manifold of dimension one).

Suppose further that Y is compact, then since $H^o(X, \pi, F) = H^o(Y, p_*^{\pi}(F))$, it follows that $H^o(X, \pi, F)$ is finite dimensional.

We say that a π-vector bundle E on X is π-indecomposable if whenever $E = E_1 \otimes E_2$ as π-bundles, it follows that $E \cong E_1$ or E_2 (as π-bundles).

Every π-vector bundle E on X can be written as a direct sum of indecomposable π-vector bundles E_i.

If moreover $Y = X/\pi$ is compact then the E_i as well the "multiplicity" with which E_i occurs in E are determined uniquely (the proof is exactly the same as in the case of vector bundles on a compact complex manifold.

If A is the ring of π-endomorphism of E, then

$A = H^o(X, \pi, E^* \otimes E)$, ($E^*$ dual of E) is finite dimensional.

In particular, A is artinian.

A decomposition of E into π-indecomposable components is equivalent to a decomposition of the identity element of A into mutually orthogonal "indecomposable" idempotents.

We get the proof by applying the usual Krull-Remak-Schmidt theorem to A considered as a module over itself) .

Let $\mathfrak{m}_X(G)$ be the sheaf of germs of meromorphic maps of X into a complex linear group G (for example if G = GL(n), a meromorphic map is a matrix whose entries are meromorphic functions on X such that on a dense open subset of X, it defines a holomorphic map into G).

Now $\mathcal{O}_X(G)$ is a subsheaf of groups of $\mathfrak{m}_X(G)$.

The quotient sheaf of sets $\mathcal{O}_X(G) \backslash \mathfrak{m}_X(G)$ (operation on the left by $\mathcal{O}_X(G)$) is called the sheaf of germs of divisors with values in G (or G-divisors) and denoted by $D_X(G)$.

Now π operates on $D_X(G)$ and so $D_X(G)$ becomes a π-sheaf.

A π-invariant section of $D_X(G)$ is called a (π -G) divisor.

Given a (π -G) divisor Θ and a point $x \in X$, $\exists$ a π-invariant open subset U containing x such that $\Theta | U$ which is a π-invariant section of

$\mathcal{O}_X(G) \backslash \mathfrak{m}_X(G)$ restricted to U comes from a section of $\mathfrak{m}_X(G) | U$ i.e. a meromorphic map of U into G (this need not be π-invariant).

From this it follows easily that a (π -G) divisor can be defined by a datum: an open covering $\{u_i\}$ of X by π-invariant open subsets and $\{f_i\}$, where f_i is a meromorphic map of U_i into

G such that $f_i(s^{-1}x) = \lambda_i^s(x)\, f_i(x)$, $\forall\, s \in \pi$, $\lambda_i^s(x)$ being a holomorphic map of U_i into G.

Now the transition functions $f_{ij}(x) = f_i(x) f_j^{-1}(x)$ which are holomorphic maps of $U_i \cap U_j$ into G define a G-principal bundle on X and through $\lambda_i^s(x)$ we can define an operation of π on this bundle.

Thus to a (π -G) divisor we can associate a π -G bundle (which is determined only up to a π -isomorphism).

The group of π -invariant meromorphic maps of X into G operates on the set of (π -G) divisor on the right and we say that two (π -G) divisor Θ_1, Θ_2 are equivalent ($\Theta_1 \sim \Theta_2$) if $\exists$ a π -invariant meromorphic map F of X into G such that $\Theta_1 F = \Theta_2$, and it is seen that the (π -G) bundles defined by Θ_1 and Θ_2 are π -isomorphic if and only if $\Theta_1 \sim \Theta_2$.

If P is a (π -G) bundle associated to a (π -G) divisor Θ, then we see that Θ can be identified with a π -invariant meromorphic section of P.

Suppose now that G = GL(n) and that E is the associated vector bundle to P.

Then Θ is defined by n π -invariant meromorphic sections of E which are holomorphic and linearly independent in a dense open subset of X.

In general such sections need not exist and a (π -G) bun= dle need not be defined by a divisor.

Let A be the abelian category of $\mathcal{O}_X$-modules.

It has sufficiently many injectives.

Let $\Gamma_X^G : A \longrightarrow$ category of abelian groups, be the functor $F \longmapsto H^0(X, \pi, F)$ $F \in A$.

Then Grothendieck's theory applies and we define

$H^n(X, \pi, F) = R^n \Gamma^G_X$ (nth right derived functor of Γ^G_X).

Then given a short exact sequence in A, we get the familiar long exact sequence involving $H^n(X, \pi, F)$. We have indeed that $H^n(X, \pi, F)$ is isomorphic to $H^n(Y, p^{\pi}_*(F))$.

For all these questions see Chap. V, [5]

2. π-vector bundles in the case of manifolds of dimension one.

We shall suppose hereafter that X is a connected complex manifold of dimension one that π operates faithfully on X and that X as well as X mod π are Hausdorff.

Then π_x is a cyclic group, say of order n_x and we can take U_x to be isomorphic to a disc $D = \{z \mid |z| < r\}$ such that the operation of π_x on U_x is defined as : $\alpha \circ z = \zeta z$ where, ζ is a primitive n_xth root of unity.

Then X mod π = Y is a manifold of dimension one.

If $x_1, x_2 \in X$ such that $p(x_1) = p(x_2)$, then π_{x_1} and π_{x_2} are conjugate subgroups of π and therefore the function $x \longmapsto n_x$ is π-invariant and therefore gives rise to a positive integral valued function $y \longmapsto n_y$ on Y.

Now $n_y = 1$ for all but a discrete subset of Y.

Such a function is called a signature on Y and the above function is called the signature of $p:X \longrightarrow Y$.

The points $x \in X$ (resp. $y \in Y$) such that $n_x > 1$ (resp. $n_y > 1$) are called the ramification points in X (resp. Y) of $p:X \longrightarrow Y$.

Given a signature $\{n_y\}$ on Y, it can be shown that there exists always a $p: X \longrightarrow Y$ as above with signature $\{n_y\}$ and in fact that there is a unique (upto Y-isomorphism) "maximal" simply connected one with signature $\{n_y\}$, except for the case when Y is the Riemann sphere and $Y_o = \{y \mid n_y > 1\}$ reduces to one point or two points y_1, y_2 with $n_{y1} \neq n_{y2}$.

If $\mathfrak{F}$ is a coherent π - $\mathcal{O}_X$-module on X, locally free of rank n, then $p_*^{\pi}(\mathfrak{F})$ is a coherent $\mathcal{O}_Y$-module locally free of rank n (that $p_*^{\pi}(\mathfrak{F})$ is locally free is immediate since it is without torsion and coherent. Further it is seen immediately that $p_*^{\pi}(\mathfrak{F})$ is of rank n in $Y - Y_o$, where $Y_o = \{y \mid n_y > 1\}$. This implies that $p_*^{\pi}(\mathfrak{F})$ is of rank n everywhere). It is well-known that we can find n meromorphic sections of $p_*^{\pi}(\mathfrak{F})$ which are holomorphic and linearly independent at least at one point of Y.

Thus if F denotes the π-vector bundle on X defined by $\mathfrak{F}$, we conclude that F can be defined by a (π-GL(n)) divisor (a π-GL(n) divisor will be called for shortness a π-divisor).

Thus every π-vector bundle on X can be defined by a π-divisor.

Let $\mathfrak{F}$ be a coherent $\pi - \mathcal{O}_X$-module.

Then we have $H^i(X, \pi, \mathfrak{F}) = 0$ for $i \geq 2$, since we have $H^i(X, \pi, \mathfrak{F}) = H^i(Y, p_*^{\pi}(\mathfrak{F}))$ and $p_*^{\pi}(\mathfrak{F})$ is a coherent $\mathcal{O}_Y$-module.

Proposition 2. Given a π-vector bundle E on X (of rank n) and $x \in X$, a π_x-invariant open neighbourhood U_x of X such that E/U_x can be defined by a representation $\rho: \pi_x \longrightarrow GL(n)$ and ρ is determined uniquely in its equivalence class of representations (i.e. locally at x, π-vector bundles are classified by representations of π_x into GL(n)).

Proof. Let $A = (\mathcal{O}_X(G))_x$, $G = GL(n)$.

Now E is determined lically at x by an element of $H^1(\pi_x, A)$.

We have a canonical map

$$\chi : H^1(\pi_x, GL(n)) \longrightarrow H^1(\pi_x, A).$$

We have to show that χ is bijective.

Given a cocycle $\alpha \longrightarrow f_\alpha$, $f_\alpha \in (\mathcal{O}_X(G))_x$ evaluation at x i.e. taking the value of f_α at x defines a map

$$j: H^1(\pi_x, A) \longrightarrow H^1(\pi_x, GL(n)).$$

We see that $j \circ \chi$ = Identity.

Thus is suffices to show that χ is surjective.

Now E is defined locally by a π-divisor.

This means that if $\alpha \longrightarrow f_\alpha(x)$ is a 1-cocycle represen= ting an element of $H^1(\pi, A)$, $\exists\ \Theta \in GL(n, K_x)$, K_x = quotient field of $\mathcal{O}_{X,x}$ such that

$$\Theta(\alpha \circ z) = f_\alpha(z)\,\Theta(z), \quad \alpha \in \pi_x, \ z \text{ a neighbourhood of } x.$$

Let $\varphi_\alpha = f_\alpha(x)$; then φ_α defines a representation of π_x into GL(n). Set

$$\Psi(z) = \sum_{\alpha \in \pi_x} \varphi_\alpha \Theta(\alpha^{-1} z).$$

We have

$$\Psi(\beta z) = \sum_{\alpha \in \pi_x} \varphi_\alpha \Theta(\alpha^{-1} \beta z) = \sum_{\alpha \in \pi_x} \varphi_\beta \varphi_{\gamma^{-1}} \Theta(\gamma z)$$

(setting $\alpha^{-1}\beta = \gamma$)

$$= \varphi_\beta \sum \varphi_{\gamma^{-1}} \Theta(\gamma z) = \varphi_\beta \cdot \Psi(z).$$

i.e. $\Psi(\beta z) = \varphi_\beta \Psi(z)$.

Further, we have

$\Psi(z) \quad \Theta^{-1}(z) = \sum_{\alpha \in \pi_x} \varphi_\alpha \;(z)\; f_{\alpha^{-1}}(z)$ since

$\Theta(\alpha^{-1} z) = f_{\alpha^{-1}}(z) \, \Theta\, (z)$.

Since $f_{\alpha^{-1}}(x) = \varphi\,(x)$, we get $\Psi(x) \quad \Theta^{-1}(x) = n_x$. Id.

This shows that $\Psi(z) \quad \Theta^{-1}(z)$ defines an element of $(\mathcal{O}_X(G))_x$, which in turn implies that $\Psi(z)$ and $\Theta(z)$ define the same divisor locally at x.

But the π_x-GL(n) bundle defined by Ψ is defined by the representation φ_α and this proves that χ is surjective.

This proves the proposition.

Remark 1. Let E be a π-vector bundle on X of rank r and $x_1, x_2 \in X$ such that $p(x_1) = p(x_2)$.

Now π_{x_1} and π_{x_2} are conjugate subgroups of π; choose an isomorphism of π_{x_2} onto π_{x_1} by one such conjugation.

Then by the above proposition $\hat{E}$ is defined locally at x_1 and x_2 by conjugate (or equivalent) representations of π_{x_1}.

We say that E is locally of type τ, where τ represents representa= tions $\rho_i : \pi_{x_i} \longrightarrow GL(r)$, x_i being a point of X chosen over every ramification point y_i Y of $p: X \longrightarrow Y$ if at x_i, E is locally π_{x_i} -isomorphic to the π_{x_i} - vector bundle defined by ρ_i.

All π-vector bundles of the same local type τ are mutually locally isomorphic (at every point of X).

Remark 2. The above proposition remains valid even in the case when dim X > 1.

For the above proof to go through, we have only to show that every π-vector bundle E on X (of rank r) can be defined locally by a π-divisor.

For this we note that for a suitably chosen π_x invariant neighbourhood U_x of x, $V = U_{x/\pi_x}$ is a Stein space.

Let E be the coherent sheaf associated to E; then one see that the coherent sheaf $p_*^{\pi}(E) = F$ is locally free of rank r outside an analytic subset of V.

Then from the fact that V is a Stein space, one concludes that F has n-sections which are linearly independent in a non-empty open subset of V.

This implies that E can be defined locally by a π-divisor.

Another way of proving this is to show that $H^1(\pi_x, K_x)$ is trivial (K_x quotient field of $\mathcal{O}_{X,x}$), which is again well-known.

Remark 3. Let $x \in X$, then $H^1(\pi_x, GL(r))$ which is the equivalence class of representations of π_x into GL(r) can be identified with the set of all diagonal representations (π_x being cyclic of finite order) ρ of the form

$$\rho(\alpha) = \begin{pmatrix} \zeta^{d} & & 0 \\ & \ddots & \\ 0 & & \zeta^{d_r} \end{pmatrix}$$

where α is a generator of π_x, ζ is the primitive n_xth root of unity defined by $\alpha \cdot z = \zeta \cdot z$ (z local coordinate at x) and $0 \leq d_1 \leq \ldots \leq d_r < n_x - 1$

Given a π-bundle E of rank r on X if $\rho : \pi_x \longrightarrow GL(r)$ is the representation defining E locally at x, we see that if we nor= malize d_i as above, the d_i's are well-determined (i.e. independent of the local coordinates as well as the choice of α).

We see that if Δ is the matrix

$$\Delta = \begin{pmatrix} z^{d} & & 0 \\ & \cdot & \\ 0 & & \cdot z^{d_r} \end{pmatrix}, z \text{ a local coordinate at } x$$

then every π-divisor defining Θ is of the form

$$\Delta \cdot \Theta_o$$

where Θ_o is invariant under π_x i.e. entries in Θ_o are meromor= phic functions in $w = z^{n_x}$, which can be identified with a local coor- dinate at $y = p(x)$.

<u>Remark 4.</u> A π-divisor associated to a π-vector bundle E on X can be described purely in terms of Y. For every y, let A_y denote $\mathcal{O}_{Y,y}$ if $n_y = 0$ and if $n_y > 1$ denote the power series ring in $w^{\frac{1}{n_y}}$, w a local coordinate at y.

Let L_y denote the quotient field of A_y.

Then a π-divisor associated to E is a map $y \longmapsto [\Theta_y]$, where $\Theta_y \in GL(r, L_y)$ and $[\Theta_y]$ denotes the coset in $GL(r, A_y) \diagdown^{GL(r, L_y)}$ determined by Θ_y such that

(i) for all but a discrete subset of Y, Θ_y GL(r, A_y)

(ii) if $n_y > 1$,

$$\Theta_y = \begin{pmatrix} w^{d_1/n_y} & & 0 \\ & \ddots & \\ 0 & & w^{d_r/n_y} \end{pmatrix} (\Theta_o)_y, \quad 0 \quad d_1 \leq \cdots \leq d_n < n_y .$$

where $\Theta_o \in GL(r, K_y)$ - K_y-quotient field of $\mathcal{O}_{Y,y}$.

The π-divisor is denoted often by Θ for shortness.

We recall that if Θ_1, Θ_2 represent π-divisors asso= ciated to π-bundles E_1 and E_2 of rank r then E_1 and E_2 are isomorphic if and only if there exists an $F \in GL(r, M)$-M field of meromorphic functions on Y such that $\Theta_1 . F = \Theta_2$.

Remark 5 . Let Θ be a π-divisor on Y representing a π-vector bundle E of rank r on X as above.

Then a π-invariant section of E on $p^{-1}(U)$, U open in Y can be identified with a column matrix $\mathfrak{f} = \begin{pmatrix} f_1 \\ \vdots \\ f_1 \end{pmatrix}$ such that the entries in $\Theta_y \mathfrak{f}$ are in A_y for every $y \in U$. If y is such that $n_y > 1$, we see that

$$\Theta_y \mathfrak{f} \text{ has elements in } A_y \Longleftrightarrow (\Theta_o)_y . \mathfrak{f} \text{ has elements}$$

in $\mathcal{O}_{Y,y}$.

From this we conclude immediately that the vector bundle p_*^{π} (E) on Y can be defined by the GL(r)-divisor Φ such that

$$\Phi_y = \Theta_y \text{ if } n_y = 1 \text{ and } \Phi_y = (\Theta_o)_y \text{ if } n_y > 1.$$

We now define the π-degree of a π-divisor Θ in the ca=

se Y is compact as follows:

$$\pi\text{-}\deg \Theta = \sum_{y \in Y} \text{order } (\det \Theta_y)$$

($\det \Theta_y \in L_y$ and for an element $f \in L_y$, $f \neq 0$, order f denotes the rational number $\frac{p}{n_y}$, where p is the multiplicity of zero or pole of f at 0 considered as a function of $z = w^{\frac{1}{n_y}}$).

We define the degree of a π-vector bundle E as the degree of a π-divisor Θ on Y representing E.

We see that this is well defined.

Taking Θ as above, we find

$$\pi\text{-}\deg E = \deg p_*^{\pi}(E) + \sum_{y, n_y > 1} \frac{d_1 \ldots d_r}{n_y}, \quad (d_i \text{ depend on } y).$$

$$= \pi\text{-}\deg \overset{r}{\Lambda} E.$$

We note that if π is a finite group i.e. when X is compact,

$$\pi\text{-}\deg E = \frac{1}{\text{ord } \pi} \cdot \deg E$$

(where $\deg E$ = degree of the line bundle $\overset{r}{\Lambda} E$).

Proposition 3. Let E be a (π-G) bundle defined by a representation $\rho : \pi \longrightarrow GL(r) = G$ and Y be compact. Then

$$\pi\text{-}\deg E = 0 .$$

Proof. $\exists$ a π-invariant meromorphic section of E i.e. π a meromorphic map $F : X \longrightarrow GL(r)$ such that

$$F(\alpha \cdot z) = \rho(\alpha) F(z),$$

and F gives rise to a divisor defining E. Let $f = \det F$. Then $f(\alpha \cdot z) = (\det \rho(\alpha)) \cdot f(z)$. Now

$$g = df/f$$

defines a π-invariant meromorphic differential on X and therefore defines canonically a meromorphie differential g' on Y.

Now the sum of the residues at the pales of g' is zero.

Now one checks that the residue of g' at y is precisely the order of f i.e. order of det F at y in the sense defined above or more precisely the order of the divisor defined by Ⓗ at y.

We conclude then, that

$$\sum_{y \in Y} \text{Order of } F \text{ at } y = 0 .$$

This proves the proposition.

We have now the basic

Theorem 1 (Weil) . Let X, Y be as above and suppose further that Y is compact and X simply connected. Let E be a π-vector bundle on X and E_i its π-indecomposable components.

Then E is defined by a linear representation of π if and only if π-deg $E_i = 0 \; \forall i$.

We will not prove this theorem here.

For proof cf. [19] or [4] .

Let K_X denote the line bundle associated to the sheaf of germs of holomorphic 1-forms on X.

Then K_X is canonically a π-line bundle. Then we have

Theorem 2 . (Duality theorem). Let E be a (holomorphic)

π-vector bundle on X and E^* denotes its dual π-vector bundle.

Suppose that Y is compact.

Then $H^1(X, \pi, E)$ can be identified canonically with the dual of the finite dimensional vector space $H^0(X, \pi, E^* \otimes K_X)$.

Proof. This is an immediate consequence of the usual duality theorem (on Y) because of

(i) $p_*^{\pi}(K_X) = K_Y, \quad p_*^{\pi}(E^* \otimes K_X) = (p_*^{\pi}(E))^* \otimes K_Y$

where K_Y denotes the line bundle on Y associated to the sheaf of germs holomorphic 1-forms on Y and

(ii) $\qquad H^i(X, \pi, F) = H^i(y, p_*^{\pi}(F))$

where F is a coherent π-sheaf on X (we have mentioned (ii) towards the end § 1).

Let for every $y \in Y$, τ_y represent a character of π_x for some fixed choice of $x \in X$ such that $p(x) = y$ (the only non-trivial case is when $n_y > 1$) and $\tau = \{\tau_y\}_{y \in Y}$.

We say that a π-vector bundle E of rank r on X is τ-special, if for every $y \in Y$, E is defined in a neighbourhood of x, where x is the point chosen over y, by the representation $\rho : \pi_x \longrightarrow GL(r)$, $\rho = \tau_y$.

Id (Id -identity element of $GL(r)$).

Let $\mathcal{B}(\tau)$ represent the category of τ-special vector bundles, being fixed.

Then we have

Proposition 4. Let $E_1, E_2 \in \mathcal{B}(\tau)$.

Then the canonical homomorphism

$$\chi : \text{Hom}\ (E_1, E_2) \longrightarrow \text{Hom}\ (p_*^{\pi}\ (E_1),\ p_*^{\pi}\ (E_2))$$

is an isomorphism (i.e. the functor $E \longrightarrow p_*^{\pi}$ (E) from the category $\mathcal{B}(\tau)$ to the category of vector bundles on Y is fully faithful).

Proof. That χ is injective is immediate.

We have only to show that χ is surjective.

Given $y \in Y$, choise $x \in X$ over y given by the definition of τ. Choose a π_x-invariant neighbourhood of x such that the operation of π_x on U_x is given by $\alpha \cdot z = \zeta z$, where ζ is an n_xth root of 1 (n_x-the integral valued function on X defined before) and α is a generator of π_x.

Then for $\tau_y : \pi_x \longrightarrow \mathcal{C}$, we have $\tau_y(\alpha) = \zeta^d$ for some $d \geqslant 0$.

If $E_1 \in \mathcal{B}(\tau)$, the associated coherent sheaf to E_1 can be represented as a free module M over $\mathcal{O}_{X,x}$ with basis say $e_1, \dots, e_r$ such that if $m \in M$, $m =$

$$\sum_{i=1}^{r} f_i(z)\, e_i,\ f_i \in \mathcal{O}_{X,x}$$

$$\alpha \cdot m = \sum_{i=1}^{r} \zeta^{-d} f_i(\zeta z).$$

One finds that $m \in M^{\pi}$, $m = \sum_i z^d f_i(z)\, e_i,\ f_i \in \mathcal{O}_{Y,y}$.

This implies that the $\mathcal{O}_{X,x}$ submodule of M generated by M is $z^d.M$.

Let N be a free $\mathcal{O}_{X,x}$ module representing E_2 in the same manner as M_1 represents E_1.

Then we find that an element of $\mathrm{Hom}_{\mathcal{O}_{Y,y}}(M, N)$ extendes to a $\mathcal{O}_{X,x}$ homomorphism of the submodule $z^d M$ into the submodule $z^d N$ i.e. in fact it extends to a $\mathcal{O}_{X,x}$ homomorphism of M into N.

This is obviously π-invariant. This shows that an element of $\mathrm{Hom}(p_*^{\pi}(E_1), p_*^{\pi}(E_2))$ extends to a π-invariant homomorfhism of E_1 into E_2 and the surjectivity of χ follows.

This proves the proposition.

Proposition 5. Let Y be a compact Riemann surface with genus $g \geq 1$ and y_o a fixed point of Y.

Let $\{n_y\}$ be the signature on Y such that $n_{y_o} = n$ and $n_y = 1$ for $y \neq y_o$.

Let $p : X \longrightarrow Y$ be the simply connected Riemann surface with signature $\{n_y\}$.

Let x_o be a point over y_o and τ the character of π_{x_o} such that $\tau(\alpha) = \zeta^q$, $n > q \geq 0$, α a generator of π_{x_o} such that $\alpha \cdot z = \zeta \cdot z$, z local coordinate at x_o.

Let F be a vector bundle on Y of rank n and degree-q.

Then $\exists$ a τ-special bundle E on X such that (i) $p_*^{\pi}(E) = F$ and (ii) E is associated to a representation of π.

Proof. Let Φ be a divisor $y \longrightarrow \Phi_y$, $y \in Y$ defining F.

We now define a π-divisor Θ by defining $\Theta_y = \Phi_y$, $y \neq y_o$, $y \in Y$ and $\Theta_y = \Delta \cdot \Phi_y$ where $\Delta = z^{q/z}$. Id. (Id - Identity matrix of order n).

Let E be a (π -G) bundle defined by (H) . Then as we have seen before p_*^{π} (E) is isomorphic to F (cf. Remark 5, Prop. 2, Chap. I).

Further E is indecomposable since F is so. We see that

$$\pi\text{- deg } E = \text{deg. } F - \frac{1}{n} \cdot nq = 0$$

Therefore by Theorem 1, we conclude that E is associated to a representation of π , q. e. d.

Remark. Given a vector bundle F on a compact Riemann surface Y of rank n, we can find a line L such that

$$-n < \deg (F \otimes L) \leq 0.$$

Thus in view of Prop. 4 and 5, the study of vector bundles on Y of arbitrary degree can be reduce to the study of π -bundles defined by representations.

If the degree is divisible by the rank, the theory of π -bundles when π -operates freely suffices but otherwise we need the case when π does not operates freely.

Proposition 6. Let X be a simply connected Riemann surface and a (faithful) discontinuous group of automorphisms of X.

Then $\exists$ a normal subgroup π_0 of π and of finite index in π such that π_0 operates freely on X.

Proof. It is classical that X is either the Riemann sphere, the plane or the upper half plane.

In all these cases the group of automorphisms of X is a group of matrices.

The proposition is now an immediate consequence of the following

Lemma 2 (Selberg) . Let M be a finitely generated group

of matrices.

Then $\exists$ a normal subgroup M_o of M of finite index such that M_o does not contain any element of finite order.

For proof see [13]

Remark . Let X be as in Prop. 6 and π a discontinuous group of automorphisms such that $Y = X \bmod \pi$ is compact (Hausdorff).

Chooose π_o as in the proposition and let $X_1 = X \bmod \pi_o$.

Let $p: X \longrightarrow y$, $q: X \longrightarrow X_1$ and $p_1: X \longrightarrow Y$ be the canonical maps.

Let E be a π-vector bundle on X.

Then $q_*^{\pi_o}(E)$ has a natural structure of Γ-bundle where $\Gamma = \pi/\pi_o$.

Now $q_*^{\pi_o}$ is an equivalence of categories since π_o operates freely.

This implies that the study of π-bundles on X is reduced to the study of Γ-bundles on X_1.

But now X_1 is a compact Riemann surface and Γ is a finite group.

Now X_1 and Y have natural structures of algebraic schemes (smooth and projective over $\mathbb{C}$) and Γ is a group of automorphisms of this algebraic structure.

Besides a holomorphic Γ- vector bundle on X_1 becomes an algebraic Γ-vector bundle for this structure and the algebraic classification and the holomorphic classification of these bundles coincide (cf [15]).

This reduces the study of π -bundles on X to an algebraic problem.

3. <u>Manifold of irreducible unitary representations of</u>

Let Y be a compact Riemann surface. Let $\{n_y\}$ be a signature on Y and $\{y_i\}$, i = 1,.... m the points of Y where $n_y > 1$.

Let $n_{y_i} = n_i$.

Let $p:X \longrightarrow Y$ be the simply connected Riemann surface with the signature n_y (we suppose that we have chosen the signature such that X exists) and π the discontinuous group operating on X such that $Y = X/\pi$.

Let g be the genus of Y.

Then it is a classical result that can be identified with the group on the letters $A_1, B_1, \dots, A_g, B_g, C_1, \dots, C_m$ subjected to the relations

$$A_1 B_1 A_1^{-1} B_1^{-1} \dots A_g B_g A_g^{-1} B_g^{-1} C_1 \dots C_m = \mathrm{Id}.$$

$$C_1^{n_1} = C_2^{n_2} = \dots = C_m^{n_m} = \mathrm{Id}.$$

<u>Proposition 7.</u> Let ρ be a representation on a vector space V (over $\mathbb{R}$) such that d = dim V such that ρ is unitary (or more generally leaving invariant a non-degenerate bilinear form on V).

Then we have

$$\dim_{\mathbb{R}} H^1(\pi, \rho) = 2d\,(g-1) + 2\dim_{\mathbb{R}} H^0(\pi, \rho) + \sum_{\nu=1}^{m} e_\nu$$

where e_ν is the rank of the endomorphism $(\mathrm{Id} - \rho(C_\nu))$ of V.

($H^i(\pi, \rho)$ denotes the ith cohomology group of π in V for the action through ρ).

<u>Proof.</u> Let us first indicate a proof for the case when the signature is trivial i.e. π operates freely on X (cf. [11]).

Then the representation $\rho : \pi \longrightarrow \mathrm{Aut}\ V$ defines a <u>local system</u> L on Y (i.e. a locally constant sheaf) of d dimensional vector spaces.

Then we have

$$\dim_{\mathcal{R}} H^o(Y,L) - \dim_{\mathcal{R}} H^1(Y,L) \quad \dim_{\mathcal{R}} H^2(Y,L) = 2d(1-g),$$

(similar to the usual Euler-Poincaré formula).

We have isomorphisms $H^i(Y,L) \longrightarrow H^i(\pi, \rho)$, $0 \leq i \leq 1$ (since the universal covering of Y i.e. X is a disc, we have indeed isomorphisms $H^i(X,L) \longrightarrow H^i(\pi, \rho)\ \forall i$ cf. § 9, Ch. XVI, [3]).

Now L is isomorphic to its dual local system since ρ leaves invariant non-degenerate bilinear form.

Therefore by the duality theorem (Th. 1.14, [16]), we have

$\dim_{\mathcal{R}} H^o(X,L) = \dim_{\mathcal{R}} H^2(X,L)$. Thus in this case, we get

$$\dim_{\mathcal{R}} H^1(\pi, \rho) = 2d\ (g-1) \quad 2 \dim_{\mathcal{R}} H^o(\pi, \rho).$$

In the general case, a similar proof should be possible.

However an explicit proof of this proposition along different lines is found in a paper of Weil (§ 6 and § 7, [20]).

<u>Remark.</u> Let U(r) (here we change the notation of the Introduction) be the group of unitary matrices of rank r and $\mathfrak{u}(r)$ its Lie algebra namely the space of skew Hermitian matrices of rank r (it is a real vector space of dimension r^2).

If $\theta \in U(r)$, we denote by Ad θ, the adjoint transforma= tion on $\mathfrak{u}(r)$, namely if $M \in \mathfrak{u}(r)$, $M \longrightarrow \theta M \theta^{-1}$.

Let $\alpha_1, \ldots, \alpha_k$ be the multiplicities of the distinct roots of the characteristic polynomial of θ i.e. if

$$\theta = A \begin{pmatrix} e^{2\pi i d_1} & & 0 \\ & \ddots & \\ 0 & & e^{2\pi i d_r} \end{pmatrix} A^{-1}, \; 0 \leq d_1 \leq \ldots \leq d_r < 1$$

we have $d_1 = \ldots = d_{\alpha_1}, d_{\alpha_1} \neq d_{\alpha_1+1}, d_{\alpha_1+1} = \ldots =$

$= d_{\alpha_2}, d_{\alpha_2} \neq d_{\alpha_2+1}, \ldots$ etc.

Then we find

(i) The rank of (Id - Ad θ) on u(r) is

$$r^2 - (\alpha_1^2 + \ldots + \alpha_k^2) = \sum \alpha_i \alpha_j \quad \text{since } r = \sum \alpha_i.$$

(For this it suffices to make the computation for a diagonal θ).

(ii) Make U(r) operate on itself by inner conjugation.

Then the isctropy group at θ for this action is of (real) dimension $\sum \alpha_i^2$ so that the dimension of the orbit through θ is of (real) dimension $r^2 - (\sum \alpha_i^2)$.

Thus we have

(iii) Rank of (Id - As θ) on $\mathfrak{u}(r)$ = the dimension of the orbit through θ for the action of u(r) on itself by inner conjugation.

Let ρ be a representation of π into U(r). Let

$\theta_\nu = \rho(C_\nu)$, $1 \leq \nu \leq m$.

Let W_ν be the orbits through θ_ν in U(r) for the action of U(r) onto itself by inner conjugation.

Let R be the set all representations ρ of in U(r) such that $\theta \in W_\nu$, $1 \leq \nu \leq m$ (i.e. the conjugacy classe of $\rho(C_\nu)$ are fixed).

Then we can identify $\rho \in R$ with the point

$$(\rho(A_i), \rho(B_i), \rho(C_\nu)) \in U(r)^{2g} \times W, \; 1 \leq i \leq g, \; 1 \leq \nu \leq m$$

$$W = \prod_{\nu=1}^{m} W_\nu, \quad 1 \leq \nu \leq m,$$

Let $\chi : U(r)^{2g} \times W \longrightarrow U(r)$ be the real analytic map defined by $(a_1, \ldots, a_g, b_1, \ldots, b_g, c_1, \ldots, c_m)$

$$\longrightarrow a_1 b_1 a_1^{-1} b_1^{-1} \ldots a_g b_g a_g^{-1} b_g^{-1} c_1 \ldots c_m$$

Then $R = \chi^{-1}(e)$ and R is therefore a closed real analytic subset of $U(r)^{2g} \times W$.

If ρ is a representation of π into U(r), let us denote by Ad ρ the representation of π into $\mathfrak{u}(r)$ defined by

$$((\mathrm{Ad}\,\rho)(\theta))(M) = \rho(\theta) M \rho(\theta^{-1}), \quad M \in u(r), \; \theta \in \pi .$$

With the above notations, we have the following

Lemma 3. Let $\rho \in U(r)^{2g} \times W$.

Then the kernel of the differential map $d\chi$ of the map $\chi : U(r)^{2g} \times W \longrightarrow U(r)$ (defined above) can be canonically identified with the space $Z^1(\pi, \mathrm{Ad}\,\rho)$ of 1-cocycles of for the action

Ad ρ on $\mathfrak{u}(r)$.

Proof. Let $M(r)$ denote the space of all $(r \times r)$ matrices over $\mathcal{C}$

Let D be the ring of dual numbers over $\mathcal{C}$ i.e. the algebra over $\mathcal{C}$ with basis $1, \varepsilon$ and $\varepsilon^2 = 0$.

If $a \in M(r)$, a tangent vector to the complex manifold $M(r)$ at a can be identified canonically with a D-valued point of $M(r)$ i.e. an element of the form $a + \varepsilon a'$.

If $a \in GL(r)$, we write this element in the form

$$(a, \alpha) = a + \varepsilon \alpha a, \quad \alpha \in M(r)$$

(to identify a tangent vector at a to one at the identity element of $GL(r)$).

If $a \in U(r)$, then a tangent vector (a, α) at a to $M(r)$ is tangent to $U(r)$ if and only if

$$(1) \qquad (a + \varepsilon \alpha a) . (a + \varepsilon \alpha a)^* = Id$$

where $*$ denotes the conjugate transpose. Now

$$(a + \varepsilon \alpha a)(a + \varepsilon \alpha a)^* = (a + \varepsilon \alpha a)(a^* + \varepsilon a^* \alpha^*)$$

$$= (a a^* + \varepsilon \alpha a a^* + \varepsilon a a^* \alpha^*).$$

But $a a^* = Id$, so that (1) is equivalent to

$$\alpha + \alpha^* = 0$$

i.e. $\alpha \in \mathfrak{u}(r)$- space of skew hermitian matrices.

We call an element $(a, \alpha) = a + \varepsilon \alpha a$ satisfying (1), a D-valued point of $U(r)$ and denote by $U(r)(D)$ the set of such point.

We note that U(r)(D) is a group under multiplications.

Given $(a, \alpha) \in U(r)(D)$, let us denote by $T_{a,\alpha}$ the affine transformation on the real vector space $\mathfrak{u}(r)$ defined by

$$\theta \longrightarrow T_{a,\alpha}(\theta) = \mathrm{Ad}(a)\,\theta + \alpha = a\,\theta\,a^{-1} + \alpha, \quad \theta \in u(r).$$

We have

$$(a_1, \alpha_1))(a_2, \alpha_2) = (a_1 + \varepsilon\,\alpha_1 a_1)(a_2 + \varepsilon\,\alpha_2 a_2) = a_1 a_2 +$$

$$+ \varepsilon\left[\alpha_1 + a_1\,\alpha_2 a_1^{-1}\right] a_1 a_2 = a_1 a_2 + \varepsilon\left[\alpha_1 + \mathrm{Ad}(a_1)\,\alpha_2\right] a_1 a_2.$$

This gives

$$T_{(a_1, \alpha_1) \circ (a_2, \alpha_2)}\,\theta = \mathrm{Ad}(a_1\ a_2)(\theta) + \left[\alpha_1\ \ \mathrm{Ad}(a_1)\,\alpha_2\right].$$

On the other hand, we check that

$$(T_{a_1, \alpha_2} \circ T_{a_2, \alpha_2})(\theta) = \mathrm{Ad}(a_1 a_2)\,\theta + \left[\alpha_1 + \mathrm{Ad}(a_1)\,\alpha_2\right].$$

This shows that $(a, \alpha) \longrightarrow T_{a,\alpha}$ defines a homomorphism of U(r)(D) into the group $\mathrm{Af}(\mathfrak{u}(n))$ of affine transformations of $\mathfrak{u}(r)$.

We see that the kernel of this homomorphism reduces to the scalar matrices of $\mathfrak{u}(r)$ i.e. $(a, \alpha) \in U(r)(D)$ with $\alpha = 0$ and scalar.

Let us recall that if $\varphi : \Gamma \longrightarrow \mathrm{Aut}\ u(r)$ (group of vector space automorphisms) is a homomorphism and $z: \Gamma \longrightarrow \mathfrak{u}(r)$ a map, then $z \in Z^1(\Gamma\ \ \varphi)$ if and only if the map $\Gamma \longrightarrow \mathrm{Af}(u(r))$ defined by

$\gamma \longrightarrow$ the element of $\mathrm{Af}(\mathfrak{u}(r))$ defined by

$$\theta \longrightarrow \varphi(\gamma)\,\theta + z(\gamma), \qquad \theta \in \mathfrak{u}(r)$$

is a homomorphism.

Let $\rho : \pi \longrightarrow U(r)$ be a representation as in the lemma. Then ρ is represented by the point

$$(a_1, b_1, \ldots, a_g, b_g, c_1, \ldots, c_m) \in U(r)^{2g} \times W$$

$$a_i = \rho(A_i),\ b_i = \rho(B_i),\ c_\nu = (C_\nu),\ 1 \leq i \leq g,\quad 1 \leq \nu \leq m.$$

Now a tangent vector to $U(r)^{2g} \times W$ at ρ can be represented by

$$t = (a_i + \varepsilon \alpha_i a_i,\ b_i + \varepsilon \beta_i b_i, c_\nu + \varepsilon \gamma_\nu c_\nu),$$

where $a_i, b_i, c_\nu \in U(r)$; $\alpha_i, \beta_i, \gamma_\nu \in \mathfrak{u}(r)$;

$1 \leq i \leq g,\quad 1 \leq \nu \leq m,$ and

$$(c_\nu + \varepsilon \gamma_\nu c_\nu)^{n_\nu} = \mathrm{Id},\quad 1 \leq \nu \leq m.$$

Now it is immediate that t is in the kernel of $d\chi$ at the point ρ if and only if

$$(3)\qquad \prod_{i=1}^{g} (a_i, \alpha_i)(b_i, \beta_i)(a_i, \alpha_i)^{-1} (b_i, \beta_i)^{-1} \prod_{\nu=1}^{m} (c_\nu, \gamma_\nu) = \mathrm{Id}.$$

$$(c_\nu \cdot \gamma_\nu)^{n_\nu} = \mathrm{Id},\ 1 \leq \nu \leq m.$$

Now (3) holds if and only if

$$(4)\qquad \prod_{i=1}^{g} Y_{a_i, \alpha_i} T_{b_i, \beta_i} T^{-1}_{a_i, \alpha_i} T^{-1}_{b_i, \beta_i} \prod_{\nu=1}^{m} T_{c_\nu,} = \mathrm{Id},$$

$$T^{n_\nu}_{c_\nu, \gamma_\nu} = \mathrm{Id},\ 1 \leq \nu \leq m,$$

since the map $(a, \alpha) \longmapsto T_{a,\alpha}$ defines a homomorphism of $U(r)(D)$ into $Af(\mathfrak{u}(r))$ whose kernel reduces to scalars in $U(r)$.

Let F be the free group on A_i, B_i, C_ν ; $1 \leq i \leq g$, $1 \leq \nu \leq m$ and N the kernel of the canonical homomorphism $\delta : F \longrightarrow \pi$.

Let $h:F \longrightarrow Af(\mathfrak{u}(r))$ be the homomorphism defined by

$$A_i \longmapsto T_{a_i, \alpha_i}, \; B_i \longmapsto T_{b_i, \beta_i}, C_\nu \longmapsto T_{c_\nu, \gamma_\nu}, \quad 1 \leq i \leq g, 1 \leq \nu \leq m.$$

Then (4) is equivalent to saying that h is trivial on N i.e. (4) is equivalent to saying that h induces a homomorphism of π into $Af(\mathfrak{u}(r))$.

By the remark made above relating 1-coclydes with homomor= phisms into $Af(\mathfrak{u}(r))$, we see that (3) is satisfied if and only if the= re is a 1-cocycle $z: \pi \longrightarrow u(r)$ in $Z'(\pi, Ad\,\rho)$ such that

$$z(A_i) = \alpha_i, \; z(B_i) = \beta_i, \; z(C_\nu) = \gamma_\nu, \quad 1 \leq i \leq g; \; 1 \leq \nu \leq m.$$

From this the lemma follows immediately since an element $z \in Z'(\pi, Ad\,\rho)$ is uniquely determined by its values on A_i, B, C_ν ; $1 \leq i \leq g$, $1 \leq \nu \leq m$.

We have now

$$\dim_{\mathbb{R}} Z^1(\pi, Ad\,\rho) = \dim_{\mathbb{R}} H^1(\pi, Ad\,\rho) + \text{space of coboundaries}$$

$$= \dim_{\mathbb{R}} H^1(\pi, Ad\,\rho) + \left[\dim \mathfrak{u}(r) - \dim_{\mathbb{R}} H^o(\pi, Ad\,\rho)\right] .$$

Then by applying Prop. 7, and Remark following Prop. 7

$$\dim_{\mathbb{R}} Z^1(\pi, Ad\,\rho) = 2r^2(g-1) + r^2 + \dim_{\mathbb{R}} H^o(\pi, Ad\,\rho) + \sum_{\nu=1}^{m} \dim W_\nu$$

$$= (2g-1)\, r^2 + \dim_{\mathbb{R}} H^0(\pi, \mathrm{Ad}\,\rho) + \sum_{\nu=1}^{m} \dim W_\nu .$$

Now $H^0(\pi, \mathrm{Ad}\,\rho) = (u(r))^\pi$ (π -invariant elements under the adjoint representation). The scalar matrices are always in $(u(r))^\pi$ so that $\dim_{\mathbb{R}} H^0(\pi, \mathrm{Ad}\,\rho) \geq 1$.

By the semi-continuity theorem on the kernel of the differential map and the implicit function theorem, we conclude that the set of points $\rho \in R$ such that $\dim H^0(\pi, \mathrm{Ad}\,\rho) = 1$, is smooth at these points.

Now $\dim_{\mathbb{R}} H^0(\pi, \mathrm{Ad}\,\rho) = 1 \Longleftrightarrow \rho$ is irreducible (because of unitary representation). Thus we have

Theorem 3. Let $R \subset U(r)^{2g} \times W$ be identified with the set of unitary representations of π into $U(r)$ such that $\rho(C_\nu)$ varies over a fixed conjugacy class -say $W_\nu \subset U(r)$ and $W = \prod_{\nu=1}^{m} W_\nu$.

Then the subset of irreducible unitary representations R_o of R is open and if it is non-empty it is smooth of real dimension

$$(2g-1)\, r^2 + 1 + \sum_{\nu} \dim W_\nu .$$

Corollary 1. Let $U(r)$ operate on R by inner conjugation.

Then the equivalence classes of irreducible unitary representations corresponds to the quotient space $R_o/U(r)$.

Now the scalars in $U(r)$ operate trivially and $PU(r)$ - unitary projective group operates freely on R_o and therefore $R_o/PU(r) = R_o/U(r)$ has a natural structure of real analytic manifold of (real) dimension $= 2r^2(g-1) + 2 + \sum_{\nu} e_\nu$.

Corollary 2. Let E be a π-bundle associated to a unitary representation $\rho : \pi \longrightarrow U(r)$. Then we have

$$\dim_{\mathcal{R}} H^1(\pi, \text{ad } \rho) = \dim_{\mathcal{R}} H^1(X, \pi, E^* \otimes E).$$

i.e. $= 2 \dim_{\mathcal{C}} H^1(X, \pi, E^* \otimes E)$.

Proof. From Prop. 7, we get

$$\text{(i)} \quad \dim_{\mathcal{R}} H^1(\pi, \text{ad } \rho) = 2r^2(g-1) \quad 2 \dim_{\mathcal{R}} H^0(\pi, \text{ad } \rho) \quad \sum_{\nu=1}^{m} e_\nu .$$

Now we have $H^i(X, \pi, E^* \otimes E) = H^i(Y, p_*^{\pi}(E^* \otimes E))$.

The Riemann-Roch theorem gives

$$\text{(ii)} \quad \dim_{\mathcal{C}} H^0(Y, p_*^{\pi}(E^* \otimes E)) - \dim_{\mathcal{C}} H^1(Y, p_*^{\pi}(E^* \otimes E))$$
$$= \deg (p_*^{\pi}(E^* \otimes E)) - r^2(g-1).$$

We have $H^0(Y, p_*^{\pi}(E \otimes E)) = H^0(X, \pi, E^* \otimes E) =$
$= H^0(\pi, \text{ad } \rho)$ (by Prop. 1).

Therefore $\dim_{\mathcal{R}} H^0(\pi, \text{ad } \rho) = 2 \dim_{\mathcal{C}} H^0(X, \pi, E^* \otimes E)$.

Thus comparing (i) and (ii), it suffices to show that (cf. definition preceding Prop. 3)

$$\deg p_*^{\pi}(E^* \otimes E) = - \frac{1}{2} \sum_{\nu=1}^{m} e_\nu .$$

Choose points $\{x_\nu\}$, $1 \leq \nu \leq m$, over the points $\{y_\nu\}$, $1 \leq \nu \leq m$ of Y.

Then the isotropy group π_{x_ν} at x_ν for the π-action on X can be identified (not canonically) with the subgroup of π generated by C_ν and thus we can identify C_ν with a generator α of π_x so that $\rho(\alpha) = \rho(C_\nu) = \theta_\nu$.

Choose ζ an n_ν th root of unity as has been done before, we write

$$\rho(\alpha) = \begin{pmatrix} \zeta^{d_1} & & 0 \\ & \cdot & \\ & & \cdot \\ 0 & & \cdot\ \zeta^{d_r} \end{pmatrix}, \quad 0 \leq d_1^\nu \leq \dots \leq d_r^\nu < n_\nu .$$

Now the bundle $E^* \otimes E$ is defined locally at x by the representation $\mathrm{Ad}\,\rho : \pi_x \longrightarrow GL(r^2)$ defined by

$$\alpha \longmapsto (\alpha) \otimes \rho(\alpha)^{-1}.$$

Now $\rho(\alpha) \otimes \rho(\alpha)^{-1}$ is a diagonal $(r^2 \times r^2)$ matrix with elements

$\zeta^{d_i^\nu - d_j^\nu}$, $1 \leq i, j \leq r$ and we observe that if $d_i^\nu - d_j^\nu < 0$, we have

$$\zeta^{d_i^\nu - d_j^\nu} = \zeta^{n_\nu - (d_j^\nu - d_i^\nu)}.$$

Thus if we write $\mathrm{Ad}\,\rho(\alpha)$ in the canonical form i.e.

$$\mathrm{Ad}\,\rho(\alpha) = \begin{pmatrix} \zeta^{l_1} & & 0 \\ & \cdot & \\ & & \cdot \\ 0 & & \zeta^{l_{r^2}} \end{pmatrix}, \quad 0 \leq l_1 \leq \dots \leq l_{r^2} < n_\nu ,$$

we check easily that if $\alpha_1^\nu, \dots, \alpha_k^\nu$ denote the multiplicities of the distinct roots of $\rho(\alpha)$, we have

$$\frac{1}{n_\nu} \sum_{i=1}^{r^2} l_i = \frac{1}{2} \sum_{\substack{1 \leq i, j \leq k \\ i \neq j}} \alpha_i^\nu \alpha_j^\nu = \frac{1}{2} e_\nu .$$

This shows that $-\deg p_*^{\pi}(E^* \otimes E) = \frac{1}{2}\sum_{\nu=1}^{m} e_\nu$

and the corollary is proved.

Proposition 8. For the group π ad above if $g \geq 2$, then $\exists$ an irreducible unitary representation π of arbitrary rank and such that $\rho(C_\nu)$ are arbitrary unitary matrices with the condition $\det \prod_{\nu=1}^{m} \rho(C_\nu) = 1$ and $(\rho(C_\nu))^{n_\nu} = 1$ for $1 \leq \nu \leq m$ (we see that these conditions are necessary).

Proof. Given r, we can always choose two unitary matrices U_1, U_2 of rank r which form an irreducible set.

Let $\theta = \prod_{\nu=1}^{m} \rho(C_\nu)$. Then we can find two unitary matrices X, Y of rank r such that

$$U_1 U_2 U_1^{-1} U_2^{-1} X Y X^{-1} Y^{-1} = \theta ,$$

because given a unitary matrix φ of determinant one and rank r, the equation

$$X Y X^{-1} Y^{-1} = \varphi$$

is easily seen to be solvable in unitary matrices. Now the representation $\rho : \pi \longrightarrow U(r)$ defined by

$$\rho(A_1) = U_1, \rho(B_1) = U_2, \rho(A_2) = X, \rho(B_2) = Y, \rho(A_i) = \rho(B_i) = Id, 3 \leq i \leq g$$

is irreducible. This proves the proposition.

C. S. Seshadri

Chapter II

Stable bundles and unitary bundles

Notation. In the following (unless otherwise stated) X will mean a smooth projective curve over $\mathcal{C}$ and π a finite group operating faithfull on X.

Let $Y = X/\pi$ and $p: X \longrightarrow Y$ the canonical morphism.

Let g, h denote the general of X, Y respectively.

Let $\mathcal{O}_Y(1)$ denote an ample invertible sheaf on Y and $\mathcal{O}_X(1) = p^* (\mathcal{O}_Y(1))$.

Then $\mathcal{O}_X(1)$ is an ample invertible π-sheaf on X.

We denote by $\mathcal{O}_X(m)$ the invertible sheaf $\mathcal{O}_X(1)^{\otimes m}$ (m fold tensor product).

Given a coherent sheaf F on X, we denote by F(m) the sheaf $F \otimes 0_X(m)$.

If V is an algebraic vector bundle on X, we denote by V(m), the vector bundle whose sheaf of germs of sections is V(m), where V is the sheaf of section of V.

We consider only algebraic schemes over $\mathcal{C}$ (i.e. schemes of finite type over $\mathcal{C}$, terminology as in [7]) and by a point of a scheme, we mean always a closed point unless otherwise stated.

Let $p_o: \tilde{X} \longrightarrow X$ be a simply connected covering of X and $p_1 = p \circ p_o$

Then Y is a quotient of a discontinuous group Γ of (faithful) automorphisms of X and there exists a normal subgroup Γ_o of Γ such that $\pi = \Gamma / \Gamma_o$ and $\Gamma_o = \pi_1(X)$.

C. S. Seshadri

1. Category of semi-stable bundles on X.

A vector bundle V on X (assumed always to be algebraic or holomorphic) is said to be semi-stable (resp. stable) if $\forall$ sub-bundle $W (\neq 0)$ of V, we have

$$\mu(W) = \frac{\deg W}{\operatorname{rk}(W)} \leq \frac{\deg V}{\operatorname{rk} V} = \mu(V) \text{ (resp. } \frac{\deg W}{\operatorname{rk} W} < \frac{\deg V}{\operatorname{rk} V} \text{ for every}$$

proper sub-bundle of V).
(we call $\mu(W)$ the reduced degree of W).

We see by an immediate application of the Riemann-Roch theorem that V is semi-stable if and only if for every sub-bundle W of V, we have

$$\frac{\chi(W(m))}{\operatorname{rk} W} < \frac{\chi(V(m))}{\operatorname{rk} V}$$

where for example, $\chi(W(m)) = \dim H^0(W(m)) - \dim H^1(W(m))$.

Since $H^1(V(m)) = 0$ for sufficiently large m, this is equivalent to saying

$$\frac{\chi(W(m))}{\operatorname{rk} W} \leq \frac{\dim H^0(V(m))}{\operatorname{rk} V} \qquad \text{for m sufficiently large.}$$

Similary, we can express the condition of stability.

We say that a π-vector bundle V on X is π-semi-stable (or semi-stable π-bundle) if the underlying vector bundle of V is semi-stable.

A π-vector bundle V on X is said to be π-stable if V is π-semi-stable and for every proper π-sub-bundle W of V, we have $\mu(W) < \mu(V)$.

Let $\mathcal{B}$ (resp. $\mathcal{B}_\pi$) denote the category of vector bundles (resp. π-vector bundles) on X.

We denote by S (resp. S_π) the category of semi-stable vector bundles (resp. π-semi-stable vector bundles) on X of degree zero.

We denote by S_n (resp. $S_{\pi,n}$) the full subcategory of S (resp. S_π) consisting of vector bundles of rank n.

Proposition 1. The category S (resp. S_π) is abelian artinian and noetherian.

In particular, every object in S (resp. S_π) has a Jordan-Holder series and the Jordan-Holder theorem holds in S (resp. S_π).

Further if $\alpha \in \mathrm{Hom}(V, W)$ V, W in S (resp. S_π), then α is of constant rank on the fibres of V.

Proof. We make use of the following simple lemmas:

Lemma 1. Let $\alpha : V \longrightarrow W$ be a morphism in $\mathcal{B}$ (resp. $\mathcal{B}_\pi$).

Then α can be decomposed as:

$$\begin{array}{ccccccccc} 0 & \longrightarrow & V_1 & \longrightarrow & V & \xrightarrow{j} & V_2 & \longrightarrow & 0 \\ & & & & & & \downarrow \beta & & \\ 0 & \longleftarrow & W_2 & \longleftarrow & W & \xleftarrow[i]{} & W_1 & \longleftarrow & 0 \end{array}$$

where V_i, W_i are vector bundles (resp. π-vector bundles), the rows are exact sequeness of vector bundles (resp. π-vector bundles), $\alpha = i \circ \beta \circ j$ and β is a generic isomorphism i.e. β induces an isomorphism on a non-empty open subset of X.

Proof. Let V , W be the $\mathcal{O}_X$-coherent modules associated to V, W respectively and $\alpha : V \longrightarrow W$ the homomorphism associated to α .

We note that $\mathcal{O}_{X,x}$ is a principal ideal ring.

Let W_1 be the minimal $\mathcal{O}_X$ submodule of W containing $\alpha(V)$ such that $W_1 \supset \alpha(V)$ and W/W_1 is locally free (W_1 can be defined as the inverse image of the torsion part of $W/\alpha(V)$).

Now α factorises as

$$\begin{array}{ccccccccc} 0 & \longrightarrow & \text{Ker} & \longrightarrow & V & \xrightarrow{j} & \text{Im}\,\alpha & \longrightarrow & 0 \\ & & & & & & \downarrow{\scriptstyle\beta} & & \\ 0 & \longleftarrow & W/W_1 & \longleftarrow & W & \xleftarrow{i} & W_1 & \longleftarrow & 0 \end{array}$$

$\alpha = i \circ \beta \circ j$, with rows exact and all the $\mathcal{O}_X$-modules occurring in these terms being locally free.

The diagram replacing the locally free sheaves by the corresponding vector bundles is the required one.

Lemma 2. Let $\beta : V \longrightarrow W$ be a homomorphism of vector bundles of rank r and the degree on X.

Then if β is a generic isomorphism, it is in fact an isomorphism.

Proof. By hypothesis $\overset{r}{\wedge} \; : \; \overset{r}{\wedge} V \longrightarrow \overset{r}{\wedge} W$ is a non-zero homomorphism and it suffices to prove that $\overset{r}{\wedge}\beta$ is an isomorphism.

Thus to prove the lemma, we can suppose that they are line bundles.

Then β can be identified with a non-zero element of

$\Gamma(X, V^* \otimes W)$.

Now $V^* \otimes W$ is a line bundle of degree zero.

This implies that $V^* \otimes W$ is the trivial line bundle and β is a constant section ($\neq 0$).

This shows that β is an isomorphism.

Proof of Proposition. Take the decomposition of α as in Lemma 1, (Chap. II).

Since V is semi-stable of degree 0, deg $V_1 \leqslant 0$.

This implies deg $V_2 \geqslant 0$.

Now $\beta : V_2 \longrightarrow W_1$ is a generic isomorphism so that

deg $W_1 \geqslant$ deg V_2 (for if $r = \text{rk } V_2$ $\overset{r}{\wedge} \beta : \overset{r}{\wedge} V_2 \longrightarrow \overset{r}{\wedge} W_1$ is non-zero, and in this case the property is clear).

This implies that deg $W_1 = 0$ since $W \in \mathbf{S}$ (resp. $\mathbf{S}_\pi$).

By Lemma 2 (Chap. II), it follows that β is an isomorphism.

This proves the proposition.

Let $V \in \mathbf{S}$ (resp. $\mathbf{S}_\pi$). Let

$$V_1 \subset V_2 \subset \dots \subset V_n = V$$

Be a Jordan-Holder series for V.

Then V_i/V_{i-1} is stable (resp. π-stable).

We denote by gr V (resp. gr V), the associated graded object $V_1 \oplus V_2/V_1 \oplus \dots \oplus V_n/V_{n-1}$ (determined only upto isomorphism but not as an object in V).

We call V_i/V_{i-1} the stable components (resp. π-stable components) of gr V (resp. gr_π V).

Note that the stable bundles (resp. π -stable bundles) are the <u>simple objects</u> in $\mathcal{S}$ (resp. $\mathcal{S}_\pi$).

Let $V \in \mathcal{B}_\pi$ (resp. $\mathcal{B}_\pi$) and L a line (resp. π-line) bundle.

Then $V \in \mathcal{S}$ (resp. $\mathcal{S}_\pi$) if and only if $V \otimes L \in \mathcal{S}$ (resp. $\mathcal{S}_\pi$).

<u>Proposition 2.</u> Let $V, W \in \mathcal{S}_r$ (resp. $\mathcal{S}_{\pi, r}$) with at least one of them being stable (resp. π -stable).

Then if $f: V \longrightarrow W$ is a <u>non-zero</u> morphism in $\mathcal{S}$ (resp. $\mathcal{S}_\pi$), then it is an isomorphism.

<u>Proof.</u> Suppose that V is stable.

Then V is a simple object in $\mathcal{S}$ (resp. $\mathcal{S}_\pi$).

This implies that $\operatorname{Ker} f = 0$.

Then by Prop. 1, it follows immediately that f is an isomorphism.

<u>Corollary.</u> Let V be a stable (resp. π -stable) bundle.

Then $\operatorname{End} V = H^o(X, V^* \otimes V)$ (resp. $\operatorname{End} V = H^o(X, \pi, E^* \otimes E)$) reduces to scalars.

<u>Proof.</u> Let $A = \operatorname{End}_\pi V$ (resp. $\operatorname{End} V$).

Then A is a finite dimensional C-algebra.

By Prop. 2 (Chap. II), every non-zero element is a unit.

This implies that $A = C$ and the Corollary is proved.

<u>Proposition 3.</u> For the category $\mathcal{S}_r$, $\exists$ an integer m_o such that for $m \geq m_o$, we have

(1) $H^1(V(m)) = 0$ for $m \geq m_o$ (by the Riemann-Roch theorem, this implies that $\dim H^o(V(m))$ is independent of $V \in \mathcal{S}_r$ for $m \geq m_o$).

(2) the canonical homomorphism $E \to V(m)$, where E represents the trivial bundle $X \times H^{o}$ $(V(m))$ is surjective i.e. $V(m)$ is a quotient bundle of E.

Proof. We note that $V \in \mathcal{B}$ is stable if and only if V^{*} (dual of V) is stable.

We shall now prove the proposition more generally for S_{r}^{α}- which denotes the category of semi-stable bundles of rank r and reduced degree α.

Let us first prove (1).

If V is semi-stable such that $\mu(V) = \alpha$, then $H^{o}(V) = (0)$ if $\alpha < 0$, for if $\exists$ $s \in H^{o}(V)$, $s \neq 0$, then s generates a line sub-bundle L of V such that $\deg L = \mu(L) \geq 0$.

The duality theorem gives that for any $V \in \mathcal{B}$,

$$\dim H^{1}(V(m)) = \dim H^{o}(V^{*}(-m) \otimes K).$$

where K is the line bundle associated to the sheaf of differentials on X.

Now $V^{*}(-m) \otimes K$ is semi- stable if V is semi-stable.

Thus $H^{1}(V(m)) = 0$ if $\mu(V^{*}(-m) \otimes K)$ 0.

Now we have

$$\mu(V^{*}(-m) \otimes K) = -\mu(V) - m \deg 0_{X}(1) + \mu(K).$$

Now $\mu(V)$ and $\mu(K)$ are fixed.

Thus for $m >> 0$, the right hand side is negative and so assertion (1) follows.

To prove (2), we proceed as follows:

Let I_{p} denote the ideal sheaf associated to a point $P \in X$

and $T_p = \mathcal{O}_X / I_p$.

Then I_p gives a line bundle of degree -1.

Because of (1) we can choose m_o such that $\forall\ m \geqslant m_o$,

$$H^1(V(m)) = 0 \text{ and } H^1(V(m) \otimes I_p) = 0 \ \forall\ P \in X.$$

Tensoring the exact sequence $0 \to I_p \to \mathcal{O}_X \to T_p \to 0$ by $\mathcal{V}(m)$ (sheaf associated to $V(m)$), we get

$$0 \longrightarrow \mathcal{V}(m) \otimes I_p \longrightarrow \mathcal{V}(m) \longrightarrow \mathcal{V}(m) \otimes T_p \longrightarrow 0 \text{ exact.}$$

Writint the cohomology exact sequence, we get

$$H^o(\mathcal{V}(m)) \longrightarrow H^o(\mathcal{V}(m) \otimes T_p) \longrightarrow 0 \text{ exact for } m \geqslant m_o.$$

This implies that the fibre of $V(m)$ at every $P \in X$ is generated by $H^o(V(m))$ and the proposition is proved.*

We say that a family of vector bundles $\{V_t\}_{t \in T}$ on X parametrized by an algebraic (resp. analytic) scheme T is <u>algebraic</u> (resp. <u>analytic)</u> if there is an algebraic (resp. analytic) vector bundle V on $X \times T$ such that $V \mid X \times t \approx V_t$.

We say that a subcategory $\mathcal{K}$ of $\mathcal{B}$ is <u>bounded</u> if there is an <u>algebraic</u> family of vector bundles $\{V_t\}_{t \in T}$ parametrized by an algebraic scheme T such that given $V \in \mathcal{K}$, there is a V_t such that $V \approx V_t$.

<u>Proposition 4.</u> Let K be a category of vector bundles

---

* This proof suggested by M. S. Naraoimhan is more direct than the one to be found in § 3, [17] .

on X of fixed rank and degree, say n and d respectively, such that it satisfies the conditions (1) and (2) of Prop. 3.

Then there is a family of vector bundles $\{V_t\}_{t \in T}$ parametrized by an (irreducible) algebraic variety T "containing" K i.e. given $W \in K$, $\exists$ $t \in T$ such that $W \approx W_t$ (in particular K is bounded).

Proof. We make use of the following well-known property (due to Serre).

Lemma 3. Let V be a vector bundle on X of rank n, $n \geq 2$ and such that $H^o(V)$ generates V.

Then $\exists$ a trivial sub-bundle I_{n-1} of V of rank (n-1) so that we have

$$0 \to I_{n-1} \to V \to V/I_{n-1} \to 0 \text{ exact and } V/I_{n-1} = \overset{n}{\wedge} V.$$

Sketch of proof of lemma. For every $P \in X$, let K_p be the kernel of the canonical homomorphism

$$H^o(V) \longrightarrow \text{Fibre of } V \text{ at } P .$$

We have to find an (n-1) dimensional linear subspace of $H^o(V)$ such that its intersection with every K_p is the linear sub-space (0).

This is done easily by counting dimensions.

Let us go to the proof of the proposition.

Let m be an integer such that $H^1(V(m)) = 0$ and $H^o(V(m))$ generates $V(m)$ $\forall$ $V \in \mathbf{K}$.

The dimension of $H^o(V(m))$ is the same for every $V \in \mathbf{K}$.

Let E be the trivial vector bundle on X of rank =

$= \dim H^0(V(m))$.

Then by Lemma 3, (Chapter II), we have

$$0 \to I_{n-1} \to V(m) \to L \to 0 \quad \text{exact } \forall\, V \in \mathbf{K}$$

or $$0 \to I_{n-1}(-m) \to V \to L(-m) \to 0 \quad \text{exact } \forall\, V \in \mathbf{K}.$$

We have $\deg V = \deg L(-m) + \deg I_{n-1}(-m)$.

It follows then that the degree of $L(-m)$ is constant say d_1 when V varies over $\mathbf{K}$.

Thus every element of $\mathbf{K}$ can be represented as an extension of a line bundle of degree d_1 by the fixed vector bundle $I_{n-1}(-m)$.

Let A denotes the affine variety on the vector space $H^1(L_1^* \otimes I_{n-1}(-m))$, where L_1 is a fixed line bundle of degree d_1 so that to each element a of A, we get a vector bundle V_a which is an extension of L_1 by $I_{n-1}(-m)$.

We check easily that $\{V_a\}_{a \in A}$ is in fact an <u>algebraic family</u> (cf. § 3, [12] for details).

We have an algebraic family $\{L_\alpha\}$ of line bundles of degree 0 parametrized by the Jacobian J of X.

Then the correspondence

$$(a, \alpha) \to V_a \otimes L_\alpha$$

defines an algebraic family of vector bundles on X parametrized by the irreducible variety $A \times J$.

This <u>contains</u> the category $\mathbf{K}$ by our construction and the proposition is proved.

Remark. The converse of the above Proposition is true, namely that if $\mathbf{K}$ is a bounded category of vector bundles on X, then we can find an integer m such that for $m \geq m_o$, $H^1(V(m)) = 0$ and $H^o(V(m))$ generates $V(m)$, $\forall\, V \in \mathbf{K}$.

When $\mathbf{K}$ reduces to one element, there are the well-known theorems of Serre (cf. [14]).

The general case follows from these theorems by an easy application of the semi-continuity theorems applied to an algebraic family containing $\mathbf{K}$ (cf. § 7, Chap. III, [7] or § 5, Chap. II, [10]).

Given a sub-category of B with bounded ranks and degrees, we see that $\mathbf{K}$ is bounded if and only there is an algebraic family of vector bundles on X containing $\mathbf{K}$.

We say that a family of π-vector bundles $\{V_t\}_{t \in T}$ on X parametrized by an algebraic (resp. analytic) scheme T is algebraic (resp. analytic), if $\exists$ an algebraic (resp. analytic) π-vector V on $X \times T$ (for the canonical action of π on $X \times T$ extending the action of π on X by taking the trivial action on T) such that $V \mid X \times t (\approx X) \approx V_t$.

We say that a category $\mathbf{K}$ of π-vector bundles on X is bounded if there is an algebraic family of π-vector bundles on X containing $\mathbf{K}$.

Proposition. 5. Let E_1, E_2 be two π-vector bundles on X of the same rank (say r), same π-degree and such that they have the same local type τ (or equivalently locally isomorphic cf. Remark 1, Prop. 2, Chap. I).

Then there is an algebraic family $\{E_t\}_{t \in T}$ of π-vector bundles parametrized by an (irreducible) algebraic variety T containing E_1 and E_2 i.e. there exist t_1, t_2 T with $E_{t_1} \approx E_1$, $E_{t_2} E_2$ (in the sense of π-isomorphism) and such that all E_t, $t \in T$, are locally of type τ.

Proof. Let $y_i \in Y$, $i = 1, \dots m$, be the points of Y over which $p : X \longrightarrow Y$ is ramified and n_i, $i = 1, \dots m$, the orders of the isotropy groups π_{x_i} at $x_i \in X$ such that $p(x_i) = y_i$.

Let z represent a local coordinate (i.e. a generator of the maximumal ideal of the algebraic local ring $\mathcal{O}_{Y,y}$) at points of Y.

Then we have matrices (cf. Remarks 4 and 5, Prop. 2, Chap. I)

$$\Delta_i = \begin{pmatrix} \frac{d^i}{n_i} \cdot & & 0 \\ & \cdot & \\ 0 & & \cdot \, d^i_r/n_i \end{pmatrix}, 0 \leq d^i_1 \leq d^i_2 \leq \dots \leq d^i_r \leq n_i, d^i_j \text{ integers}$$

such that E_1, E_2 can be defined by divisors Φ^1, Φ^2 of the form

$\Phi^k : y \longrightarrow \Phi^k_y$, $k = 1, 2$ such that

$$\Phi^k_{y_i} = z^{\Delta_i} \Theta^k_{y_i}, \quad z^{\Delta_i} = \begin{pmatrix} z^{d^i_1/n_i} & & 0 \\ & \cdot & \\ & & \cdot \\ 0 & & \cdot \, z^{d^i_r/n_i} \end{pmatrix}$$

$$\Phi^k_y = \Theta^k_y, \quad y \neq y_i$$

and $\Theta^k : y \longrightarrow \Theta^k_y$ represents a divisor for $p^{\pi}_{*}(E_k)$ $(k = 1, 2)$.

We can in fact find Φ^k so that $\Phi^k_{y_i}$ are regular maps into $GL(r)$ in some neighbourhood of y_i (for we can find rational section of the principal bundle associated to a vector bundle regular in a neighbourhood of a finite number of points of X).

Net now E be <u>any</u> π-vector bundle on X such that $p_*^\pi(E) \approx p_*^\pi(E_1)$.

Then E can be defined by a divisor Φ of the form

$$\Phi : y \longrightarrow \Phi_y ; \quad \Phi_y = \Theta_y^1 , \ y \neq y_i$$

$$\Phi_{y_i} = z^{\Delta_i} \Psi_i , \quad \Psi_i \text{ regular map of a neighbourhood}$$

of y_i into $GL(r)$.

Now by the definition of a divisor, the divisor Φ assigns to y_i the coset in $GL(r, A_{y_i}) \backslash GL(r, L_{yi})$ determined by Φ_{yi} (A_{y_i} denotes the power series ring $z^{1/n_{yi}}$ and L_{yi} is the quotient field of A_{yi}).

Consider the equation

$$(*) \qquad u.\, z^{\Delta_i} \Psi_i = z^{\Delta_i} \Psi_i' \quad \Psi_i' \in GL(r, \mathcal{O}_{Y, y_i}),$$

$$u \in GL(r, A_{yi}).$$

If $(*)$ holds, we have

$$u = z^{\Delta_i} \Psi_i' \Psi_i^{-1} z^{-\Delta_i} .$$

Set $\alpha = (\alpha_{kl}) = \Psi_i' \Psi_i^{-1}$.

Then $\alpha \in GL(r, \mathcal{O}_{Y, y_i})$.

Now (✱) holds if and only if $z^{\Delta_i} \alpha z^{-\Delta_i} \in GL(r, A_{y_i})$.

We have

$$z^{\Delta_i} \alpha z^{-\Delta_i} = (\alpha_{kl} z^{(d_k^i - d_l^i)/n_i}).$$

Now we see easily that $z^{\Delta_i} \alpha z^{-\Delta_i} \in GL(r, A_{y_i})$ if and only if

$$\alpha_{kl}(y_i) = 0 \text{ whenever } d_k^i - d_l^i < 0.$$

Thus (✱) holds if and only if the value of the matrix $\alpha = \Psi_i' \Psi_i^{-1}$ at y_i belongs to the parabolic subgroup of $GL(r)$ determined by matrices $P = (p_{kl})$ such that

$$p_{kl} = 0 \text{ whenever } d_k^i - d_l^i < 0 .$$

This shows that we can whoose $y \longrightarrow \Phi_y$ so that $\Psi_i = K_i$, where $K_i \in GL(r)$ or that

$$\Psi_i = \Theta_{y_i}^1 \; K_i \text{ (or } K_u \; \Theta_{y_i}) \text{ for some } K_i \quad GL(r).$$

i.e. by modifying $\Theta_{y_i}^1$ by a suitable constant matrix we can represent by a divisor any π-vector bundle E such that $p_*^{\pi}(E) \approx p_*^{\pi}(E_1)$.

We get thus a similar statement for π-vector bundles E such that $p_*^{\pi}(E) \approx p_*^{\pi}(E_2)$.

Let $V_1 = p_*^{\pi}(E_1)$ and $V_2 = p_*^{\pi}(E_2)$.

Then V_1, V_2 are vector bundles on Y of rank r and the

same degree.

Then by Prop. 4 (Chap. II), there is an algebraic family $\{V_t\}_{t \in T_0}$ parametrized by a quasi-projective algebraic variety T_0 such that there are two points $t_1, t_2 \in T_0$ such that $V_{t_1} \approx V_1$ and $V_{t_2} \approx V_2$.

We shall now show that there is an open neighbourhood T of T_0 containing t_1 and t_2 such that the family $\{V_t\}_{t \in T}$ can be lifted to an algebraic family $\{E_t\}_{t \in T}$ of π-vector bundles on X with $E_{t_1} \approx E_1$ and $E_{t_2} \approx E_2$ and E_t is locally of type τ for every $t \in T$.

This would prove the proposition.

Let V be a vector bundle on $Y \times T_0$ which defines the family $\{V_t\}_{t \in T_0}$

We can now find an open neighbourhood T of t_1, t_2 and an open covering $\{U_\alpha\}$ of $Y \times T$ and a meromorphic section Θ of V such that

(i) the restriction of V to U_i is trivial

(ii) $\exists$ a unique member say U_i of the covering $\{U_\alpha\}$ such that U_i contains $y_i \times T$ and U_α for $\alpha \neq i$ does not intersect $y_i \times T$.

(iii) let U_i' be the open subset in Y defined by $U_i' = pr_Y(U_i)$.

Then in $p^{-1}(U_i')$ $(p : X \longrightarrow Y)$, the function $w = z^{\frac{1}{n_i}}$ which is a local coordinate at every one of the points $p^{-1}(y_i)$ is non-vanishing except at the points $p^{-1}(y_i)$.

(iv) the section Θ is regular at the points (y_i, t) $1 \leq i \leq m$

and $\forall\, t \in \Gamma$.

Choose trivialisations of V over U_α; then the section is defined by rational maps Θ_α of U_α into $GL(r)$ such that Θ_i is regular at the points (y_i, t), $t \in T$.

Now $\Theta_{\alpha\beta} = \Theta_\alpha \Theta_\beta^{-1}$ is a regular map of $U_\alpha \cap U_\beta$ into $GL(r)$ and define transition functions for V.

We observe that the values of Θ_i at (y_i, t_k) $1 \leq i \leq m$, $1 \leq k \leq 2$ could be arbitrarily prescribed (because we can choose a rational section at y_i assuming arbitrary values at y_i).

Let $q : X \times T \longrightarrow Y \times T$ be the map $q = p$ Id and $\{W_\alpha\}$ the covering, $\{W_\alpha\} = q^{-1}(U_\alpha)$.

Then W_α is π-invariant.

Let Φ_α be the rational maps of W_α into $GL(r)$, defined by

$$\Phi_\alpha = q^{-1}(\Theta_\alpha) \quad \text{if} \quad \neq i, \quad 1 \leq i \leq m$$

$$\begin{pmatrix} w^{d_1^i} & & 0 \\ & \ddots & \\ 0 & & w^{d_r^i} \end{pmatrix} q^{-1}(\Theta_i), \quad 1 \leq i \leq m, \quad (w = z^{\frac{1}{n_2}}).$$

Now if $\Phi_{\alpha\beta} = \Phi_\alpha \Phi_\beta^{-1}$, $\Phi_{\alpha\beta}$ is a regular map of $W_{\alpha\beta} = W_\alpha \cap W_\beta$ into $GL(r)$ and define a vector bundle E on X.

By the choice of our Φ_α, π operates on E (see the discussion towards the close of § 1, Chap. I).

Now $\Phi_\alpha(t)$ (value of Φ_α at $t \in T$) is a π-divisor on Y such that its representative at y_i is of the form $z^{\Delta_i} \cdot \varphi_i$,

where φ_i is a regular map in a neighbourhood of y_i into $GL(r)$.

By the preceding discussion, by suitably choosing the values of Θ_i at $(y_i, t_k)(1 \leq i \leq m, \quad 1 \leq k \leq 2)$, we can suppose that the divisors $\Phi_\alpha(t_1)$ and $\Phi_\alpha(t_2)$ define π-vector bundles which are isomorphic to E_1 and E_2 respectively.

Now the algebraic family $\{E_t\}_{t \in T}$ of π-vector bundles gives a lifting of the family $\{V_t\}_{t \in T}$ with $E_{t_1} \approx E_1$ and $E_{t_2} \approx E_2$ and E_t locally of type τ for all $t \in T$.

This completes the proof of the proposition as remarked

Remark. Let $\{E_t\}_{t \in T}$ be an algebraic (resp. analytic) family of π-vector bundles parametrized by an algebraic (resp. analytic) scheme T.

Then the equivalence classes of the local representations associated to E_t at the ramification points of $p : X \longrightarrow Y$ are the same $\forall\, t \in T$, provided T is connected.

For this it suffices to show that, assuming $\{E_t\}_{t \in T}$ to be analytic, the above conclusion holds in a suitable neighbourhood of every point of T.

Given a point $t \in T$ and $x \in X$ which is a ramification point of $p : X \longrightarrow Y$, we can choose the open neighbourhood U_x of x (invariant by π_x) and an open neighbourhood T_o of t such that the restriction of the defining bundle E of $\{E_t\}$ to $U_x \times T_o$ is defined by a representation of π_x (cf. Remark following Prop. 2, § 2, Chap. I).

The equivalence class represented by this matrix is the local class of representations at x for every E_t, $t \in T_o$.

From this the required assertion follows.

2. Reduction to constructing orbit spaces under an algebraic group.

We say that a coherent sheaf $\mathfrak{F}$ on X has Hilbert polynomial P if

$$P(m) = \chi(\mathfrak{F}(m)) = \dim H^{0}(\mathfrak{F}(m)) - \dim H^{1}(\mathfrak{F}(m)), \text{ for every integer } m.$$

We say that a coherent sheaf $\mathfrak{F}$ on $X \times T$, T an algebraic prescheme, is flat over T if $\forall\, y \in X \times T$, $y = (x,t)$, the $\mathcal{O}_{X \times T, y}$ module $\mathfrak{F}_y$ is flat over $\mathcal{O}_{T,t}$.

Let $\mathfrak{F}_t$ denote the restriction of $\mathfrak{F}$ to $X \times t \approx X$, $t \in T$.

Then if $\mathfrak{F}$ is flat over T and T is connected, the Hilbert polynomial of $\mathfrak{F}_t$ is independent of $t \in T$.

Let now $\mathcal{E}$ be a coherent sheaf on X and P a linear polynomial with integral coefficients.

Let Quot be the functor

Quot : Algebraic preschemes $\longrightarrow$ Sets

defined by

Quot (T) = Ser of coherent sheaves F on $X \times T$ such that (i) $\mathfrak{F}$ is a quotient of $pr_X^*(\mathcal{E})$ ($pr_X : X \times T \longrightarrow X$ projection onto X) (ii) $\mathfrak{F}$ is flat over T and (iii) $\mathfrak{F}$ has hilbert polynomial P with respect to T i.e. the Hilbert polynomial of $\mathfrak{F}_t$ for every $t \in T$ is P.

Then, by Grothendieck, we know that Quot is representable by a projective algebraic scheme $Q(\mathcal{E}/P)$ (over $\mathcal{C}$) (Th. 3.2, [6]).

Because of representability, we have a uniquely determined cocherent sheaf on $X \times Q(Q = Q(\mathcal{E}/P))$ such that $\mathfrak{F}$ is flat over Q, is a quotient of $pr_X^*(\mathcal{E})$, has Hilbert polynomial P with respect to Q and is universal for all the coherent sheaves Quot (T) on $X \times T$.

Suppose now that $\mathcal{E}$ is a π-sheaf.

Then we see that π operates on Quot (T) and this operation is functorial in T.

From this, we conclude that π operates canonically as a group of automorphisms of the scheme $Q = Q(\mathcal{E}/P)$.

Let $Q^{\pi}(\mathcal{E}/P) = Q^{\pi}$ represent the canonical closed <u>subscheme of π-invariant points</u> of $Q(\mathcal{E}/P)$ defined as follows: the (geometric) points of Q^{π} are precisely the points of Q invariant under π, if $q \in Q^{\pi}$ choose an affine open subset $U = \operatorname{Spec} R$ of Q containing q and which is invariant under π; then the ideal which defines the closed subscheme $Q^{\pi} \cap U$ of U is the ideal of R generated by elements of the form $f - \alpha \cdot f$, $f \in R$ and $\alpha \in \pi$.

Now we see easily that Q^{π} represents in fact the functor

$T \longmapsto (\mathrm{Quot}\ (T))^{\pi}$ (subset of π-invariant elements of Quot (T)).

This results easily from the fact that if $Z = \operatorname{Spec} B$ is an affine scheme on which a finite group π operates and J is the ideal of B generated by elements of the form $f - \alpha \circ f$, $f \in B$ and $\alpha \in \pi$, then a homomorphism of rings $B \longrightarrow A$ is π-invariant if and only if it is zero on J.

Thus if $\underline{Z}$ is the functor

$$\underline{Z} : (\text{Algebraic Preschemes}) \longrightarrow (\text{Sets})$$

such that Z represents $\underline{Z}$ and $\underline{Z}^{\pi}$ is the functor $\underline{Z}^{\pi}(T) = (\underline{Z}(T))^{\pi}$ (π-invariant subset of $\underline{Z}(T)$), then we see that Spec (B/J) represents $\underline{Z}^{\pi}$.

We deduce the above assertion easily from this.

Now $\mathcal{G} \in (\mathrm{Quot}(T))^{\pi}$, we see that $\mathcal{G}$ is in fact a π-coherent sheaf and the canonical homomorphism

$$pr_X^*(\mathcal{E}) \longrightarrow \mathcal{G}$$

is in fact a π-homomorphism.

Thus Q^{π} represents the functor $T \rightarrow (\mathrm{Quot}\ (T))^{\pi}$ Set of all π-coherent sheaves $\mathcal{G}$ on $X \times T$ which are (i) π-quotients of E (ii) $\mathcal{G}$ is flat over T and has Hilbert polynomial P with respect to T.

We shall suppose hereafter that $\mathcal{E}$ is the sheaf of germs of sections of a π-vector bundle E of rank p such that its underlying vector bundle on X is trivial.

If we refer to E simply as a vector bundle, it is without its π-structure; when we refer to E as a π-bundle it will be done so explicitly.

Then $G = \mathrm{Aut}\ E$ (group of automorphisms of E) can be identified with the full linear group GL(p).

Let $H = \mathrm{Aut}_{\pi}\ (\mathcal{E}) = \mathrm{Aut}_{\pi}\ (E)$ (π-automorphisms of E); then H is a direct product of full linear groups; in particular it is connected and reductive.

We write aften $Q(E/P)$ instead of $\mathcal{Q}(\mathcal{E}/P)$ (similary $Q^{\pi}\ (E/P)$).

We see that G (resp. H) operates as a group of automorphisms of the scheme $Q(E/P)$ (resp. $Q^{\pi}\ (E/P)$) - this again results from the fact that G operates on Quot (T) (resp. $(\mathrm{Quot}\ (T))^{\pi}$) functorially with respect to T.

Let R_1 be the subset of $Q = Q(E/P)$, consisting of points $q \in Q$ such that $\mathcal{F}_p$ is locally free (we recall that $\mathcal{F}$ is the defining quotient coherent sheaf on $X \times Q$ of $pr_X^*(\mathcal{E})$ and $\mathcal{F}_q$ is the restriction $\mathcal{F}$ to $X \times q \approx X$).

We observe that the rank of $\mathcal{F}_q$ is the same whatever be $q \in R_1$.

Let this be r.

Let $R_1^{\pi} = R_1 \cap Q^{\pi}$.

Let R be the subset of R_1 consisting of $q \in R_1$ such that

(i) the canonical linear map $H^0(E) \longrightarrow H^0(\mathcal{F}_q)$ is an isomorphism and

(ii) $H^1(\mathcal{F}_q \doteq 0$

Let $R^{\pi} = R \cap Q^{\pi}$.

Let $R^{\tau} \subset R^{\pi}$ be the subset of R^{π} consisting of all $q \in Q^{\pi}$ such that F_q is locally of a fixed type τ i.e. the set of all $q \in R$ such that F_q is locally isomorphic (i.e. for every $x \in X$, π_x-isomorphic in a suitable π_x-invariant neighbourhood of x) to a fixed F_{q_0}, $q_0 \in R$ (cf. Remark 1, Prop. 2, Chap. I).

If $q \in R_1$ (resp. R_1^{π}) we denote by F_q the vector bundle (resp. π-vector bundle) associated to $\mathcal{F}_q$.

Proposition 6. (i) R_1 and R (resp. R_1^{π}, and R^{π}, R^{τ}) are G-invariant (resp. H-invariant) open subschemes of Q (resp. Q^{π}).

(ii) Let $q_1, q_2 \in R$ (resp. R^{π}); then F_{q_1} is isomorphic (resp. π-isomorphic) to F_{q_2} if and only if q_1, q_2 lie in the same orbit under G (resp. H), $G = \text{Aut } E$, $H = \text{Aut}_{\pi} E$.

(iii) R (resp. R^{π}) has a local universalproperty in the following sense.

Let $\{V_t\}_{t \in T}$ be an algebraic or analytic family of vector bundles (resp. π-vector bundles) on X parametrized by an algebraic scheme T such that

(a) the Hilbert polynomial of every V_t is P

(b) $H^1(V_t) = 0$ for every $t \in T$.

(c) $H^o(V_t)$ generates V_t for every $t \in T$ (resp. the π-vector bundle on X associated to the canonical π-module $H^o(V_t)$ is π-isomorphic to E and $H^o(V_t)$ generates V_t).

Then for every $t \in T$, there is an open neighbourhood T_o of t and a morphism $f : T_o \longrightarrow R$ (resp. R^{π}) such that if $q = f(t)$, F_q is isomorphic (resp. π-isomorphic) to V_t.

(iv) R (resp. R^{τ}) is smooth and irreducible.

Further R^{π} is smooth, its connected components being of the form R^{τ} for some τ.

(v) (a) If R is non-empty, then

$$\dim R = (r^2 (g-1) + 1) + (\dim G-1)$$

where r is the integer such that F_q is of rank r, $q \in R$ (g = genus of X).

(b) Let τ associate the representation ρ_ν of π_{x_ν}, $1 \leq \nu \leq m$ (cf. Remark 1, Prop. 2, Chap. I)

$$\rho_\nu(\alpha) = \begin{pmatrix} \xi^{d_1^\nu} & & 0 \\ & \ddots & \\ 0 & & \xi^{d_r^\nu} \end{pmatrix}, \quad 0 \leq d_1^\nu \leq \dots \leq d_r^\nu < n_\nu,$$

where x_ν is a point of X chosen over y_ν, $\{y_\nu\}$ being the ramification points in y of $p : X \longrightarrow Y$, α is a generator of π_{x_ν} and ξ is an n_ν-th root of unity as chosen often in § 2, Chap. I

Let $\alpha_1^\nu, \dots \alpha_{K_\nu}^\nu$ be the multiplicities of the distinct roots of the characteristic polynomial of $\rho(\alpha)$) and

$$e_\nu = \sum_{\substack{1 \leq i,\ j \leq r \\ i \neq j}} \alpha_i^\nu \alpha_j^\nu$$ (see Remark following Chap. I and Cor. 2 of Th. 2, Chap. I)

Then if R^τ is <u>non-empty,</u> we have

$$\dim R^\tau = (r^2 (h-1) + 1) + 1) + \frac{1}{2} \sum_{\nu=1}^{m} e_\nu + (\dim H - 1)$$

where h = genus of Y.

Proof. (i) To prove that R_1 and R_1^π are open in Q and Q^π respectively is suffices to show that R_1 is open in Q.

For this, it suffices to show that if U is an affine subset of X and $q \in R_1$, then there is an open subset W of Q containing q such that the restriction of $\mathcal{F}$ (defining quotient sheaf of $pr_X^*(\mathcal{E})$ on $X \times Q$) to $U \times W$ is locally free.

Let now V be an affine open subset of Q containing q, $V = \operatorname{Spec} A'$, $U \times V = \operatorname{Spec} B'$ and $\varphi' : A' \longrightarrow B'$ the homomorphism defining the canonical projection $U \times V \longrightarrow V$.

Now the restriction of $\mathcal{F}$ to $U \times V$ is defined by a noetherian B' modùle F' which is a'-flat.

Let m be the maximal ideal of A' defining q, $S = A - m$, $A = A' S^{-1}$, $B = B' S^{-1}$, $\varphi : A \longrightarrow B$ the homomorphism induced by φ' and F the B-module $F' S^{-1}$.

Now it suffices to prove that F is a free module over B, for then it is easily seen that there is an open subset W of R_1 containing q such that the restriction of F to U W is free.

Thus the above assertion is an easy consequence of

<u>Lemma 4.</u> Let $\varphi : A \longrightarrow B$ be a homomorphism of rings such that A is a local with maximal ideal m.

Suppose that for every maximal ideal n of B, $\varphi^{-1}(n) = m$

i.e. φ (m)B (which is denoted by m.B) is contained in the radical of B.

Let F be a B-module of finite type such that it is flat over A.

Then $F/_{mF}$ is free over $B/_{mB}$, then F is free over B.

Proof of lemma. Let $F^{-} = F/_{mF}$.

Let $f_1, \ldots f_r$ be a basis of $\overline{F}$ over $B/_{mB}$.

Let $f_1, \ldots, f_r$ be some elements of F which lift $f_1, \ldots, f_r$.

Let L be a free B-module of rank r and $\theta : L \longrightarrow F$ the B-homomorphism which takes a given basis of L into $f_1, \ldots, f_r$ respectively.

Since the induced homomorphism $\overline{\theta} : \overline{L} \longrightarrow \overline{F}$, where $L = L/_{mL}$ is surjective, it follows that m. $(F/\theta(L)) = 0$.

Therefore, by Nakayame, $F/\theta(L) = 0$ i.e. θ is surjective.

Let $L_1 = \operatorname{Ker} \theta$.

Then we have the exact sequence of B-modules

$$0 \longrightarrow L_1 \longrightarrow L \longrightarrow F \longrightarrow 0 .$$

Tensoring by $A/_m$ and using $\operatorname{Tor}_1^A (F, A_m) = 0$, F being A-flat, we get

$$0 \longrightarrow L_1 \otimes_A A/_m \longrightarrow L \otimes_A A/_m \longrightarrow F \otimes_A A/_m \longrightarrow 0$$

is exact.

Since $L \otimes_A A/_m \approx L/_{mL} \approx F/_{mF} \approx F \otimes A/_m$, we see that $L_1 \otimes_A A/_m = 0$ i.e. $L_1 = mL_1$

Now L is a B-module of finite type and B being noetherian with L of finite type over B, L_1 is also of finite type over B.

Therefore, by applying Nakayama once again, we deduce that $L_1 = 0$.

Thus $\theta : L \longrightarrow F$ is an isomorphism and the lemma is proved.

As remarked above, it follows now that R_1 and R_1^{π} are open in Q and Q^{π} respectively.

We shall now show that R is open in R_1.

Let $q_o \in R$.

Then by the semi-continuity theorem, there is a neighbourhood S of q_o such that $H^1(F_q) = 0$, $\forall\, q \in S$.

Further $(pr_S)_*\ (\mathfrak{F})$ is locally free on S of rank = rank E and the corresponding vector bundle is denoted by $(pr_S)_*\ (F)$.

The restriction of the canonical homomorphism $pr_X^*(E) \longrightarrow F$ to $X \times S$ gives rise to a homomorphism α of the vector bundle $(pr_S)_*\ (pr^*\ (E))$ (which is isomorphic to the trivial bundle of rank r on S) into $(pr_S)_*\ (F)$ such that α induces an isomorphism of the fibres of these vector bundles at q_o.

Therefore there is a neighbourhood S_1 of q_o, $S_1 \subset S_o$ such that α restricted to S_1 is an isomorphism.

This implies that for every $q \in S_1$, the canonical homomorphism $H^o(E) \longrightarrow H^o(F_q)$ is an isomorphism.

This shows that R^{π} is open in Q.

That R^{τ} is open is an immediate consequence of the fact that R is open.

It follows now that R^{τ} is open by the Remark following Proposition 5 (Chap. II).

The fact that R_1 and R (resp. R_1^{π} and R^{π} , R^{τ}) are G-invariant (resp. H-invariant) are immediate.

This completes the proof of (i).

(ii) Let $q_1, q_2 \in R$ (resp. R^{π}).

Suppose that q_1, q_2 lie in the same orbit under G(resp. H).

Then by the definition of the action of G (resp. H), it is immediate that F_{q_1}, F_{q_2} are isomorphic (resp. π -isomorphic).

We observe that this is so even if we take $q_1, q_2 \in Q$ (resp. Q^{π}).

Suppose on the other hand F_{q_1} is isomorphic (resp. π -isomorphic) to F_{q2}.

This gives rise to an isomorphic (resp. π -isomorphic) of $H^o(F_{q_1})$ onto $H^o(F_{q_2})$.

But now we can identify $H^o(F_{q_1})$ and $H^o(F_{q_2})$ canonically with $H^o(E)$.

This implies immediately that there is an automorphism (resp. π -automorphism) of E which takes the quotient sheaf F_{q_1} to F_{q_2} i.e. q_1, q_2 lie in the same orbit under G (resp. H.

This proves (ii).

(iii) Let V be the defining vector bundle (resp. π -vector bundle) of the family $\{V_t\}_{t \in T}$.

Consider $\mathcal{W} = (pr_T)_*(\mathcal{V})$, where $\mathcal{V}$ is the coherent sheaf associated to V.

Then by our hypothesis $\mathcal{W}$ is locally free.

Let W be the vector bundle associated to $\mathcal{W}$.

The fibre at t of W can be identified canonically with $H^o(V_t)$.

By our hypothesis on the Hilbert polynomial of V_t , it

follows further that $\dim H^0(V_t) = \dim H^0(E)$.

Given $t \in T$, there is a neighbourhood T_o of t such that W restricted to T_o is trivial.

Then $(pr_{T_o})^*(W)$ is isomorphic (resp. π-isomorphic) to $(pr_X)^*(E)$ and V can be considered as a quotient (resp. π-quotient) bundle of $pr_X^*(E)$.

By the universal property of $Q(E/P)$ (resp. Q^π) we have a morphism $f : T_o \longrightarrow Q$ such that V_t is isomorphic (resp. π-isomorphic) to F_q ($q = f(t)$ and V_t is the coherent sheaf associated to V_t).

By the definition of R (resp. R^π), we have $f(T_o) \subset R$ (resp. R^τ).

This proves (iii).

(iv) We shall first show that R (resp. R^τ) is connected. We shall suppose that R (resp. R^τ) is <u>non-empty</u>. Otherwise there is nothing to prove.

Given q_1, q_2 in R (resp. R), then by Prop. 4 (resp. Prop. 5 of Chap. II, there is an algebraic family $\{V_t\}_{t \in T}$ of vector bundles (resp. π-vector bundles) parametrized by an (irreducible) algebraic variety T and two points $t_1, t_2 \in T$ such that $V_{t_1} \approx F_{q_1}$, $V_{t_2} \approx F_{q_2}$ V_{t_1}, V_{t_2} being the coherent sheaves associated to V_{t_1}, V_{t_2} respectively.

Let T_o be the subset of T formed by points t such that (a) $H^1(V_t) = 0$ and (b) $H^0(V_t)$ generates V_t.

Then by the same type of arguments as in the proof of (i) above, T_o is open.

Now T_o contains t_1 and t_2 and T_o is irreducible.

By the universal property (iii) above, we have an open covering of non-empty subsets $\{T_i\}$ of T_o and morphism $f_i : T \longrightarrow R$ (resp. R^{τ}) such that if $q = f(t)$, then $V_t \approx F_q$.

We have $T_i \cap T_j \neq \emptyset$ and because of (ii) above, the G-saturations (resp. H saturations) of $f_i(T_i)$ and $f_j(T_j)$ have non-empty intersection for every i, j.

Now the G-saturations (resp. H-saturations) of all $f_i(T_i)$ contain q_1, q_2 and $f_i(T_i)$ is an irreducible constructible subset of R (resp. R^{τ}).

Now q_1, q_2 being arbitrary, it follows that R (resp. R^{τ}) is connected.

To conclude the proof of (iv), it suffices to show that R (resp. R^{τ} or R^{π}) is smooth.

For $q \in R$ (resp. R^{π}), let H_q be the vector bundle defined by the exact sequence.

$$0 \longrightarrow H_q \longrightarrow E \longrightarrow F_q \longrightarrow 0$$

(where F_q is the vector bundle associated to $\mathfrak{F}_q$).

Then we have the exact sequence

$$0 \longrightarrow F_q^* \longrightarrow E^* \longrightarrow H_q^* \longrightarrow 0$$

where for example by F_q^*, we mean the dual vector bundle F_q.

Tensoring this by F_q, we get the exact sequence

$$0 \longrightarrow F_q^* \otimes F_q \longrightarrow E^* \otimes F_q \longrightarrow H_q^* \otimes F_q \longrightarrow 0$$

Writing the cohomology exact sequence, we get

$$0 \to H^o(F_q^* \otimes F_q) \to H^o(E^* \otimes F_q) \to H^o(H_q^* \otimes F_q) \to H^1(F_q^* \otimes F_q)$$

$$H^1(F^* \otimes F_q) \longrightarrow H^1(H_q^* \otimes F_q) \longrightarrow H^2(F_q^* \otimes F_q).$$

(resp. the corresponding cohomology sequence in the category of π-sheaves).

We observe that $H^2(F_q^* \otimes F_q)$ (resp. $H^2(X, \pi, F_q^* \otimes F_q)) = 0$ since X is a curve (resp. since $H^2(X, \pi, F_q^* \otimes F_q) = = H^2(Y, p_* (F_q^* \otimes F_q)$ and y is a curve).

Further $H^1(E^* \otimes F_q) = 0$ for $E^* \otimes F_q$ is a direct sim of copies of the same F_q.

We have also $H^1(X, \pi, E^* \otimes F_q) = 0$ for by the duality theorem for π-bundles (Theorem 2 Chap. I), we have

$$\dim H^1(X, \pi, E^* \otimes F_q) = \dim H^o(X, \pi, E \otimes F_q^* \otimes K)$$

where K is the line bundle corresponding to the dheaf of differentials (it is canonically a π-buncle). Now

$$H^o(X, \pi, E \otimes F_q^* \otimes K) \subset H^o(X, E \otimes F_q^* \otimes K).$$

But by the usual duality theorem

$$\dim H^o(E \otimes F_q^* \otimes K) = \dim H^1(E^* \otimes F_q) = 0 \text{ by hupothesis.}$$

Thus we deduce that

$$H^1(H_q^* \otimes F_q) = 0 \quad (\text{resp. } H^1(X\ \pi, H_q^* \otimes F_q) = 0)$$

and that the following sequence is exact

(A) $\quad 0 \to H^o(F_q^* \otimes F_q) \to H^o(E^* \otimes F_q) \longrightarrow H^o(H_q^* \otimes F_q) \to H^1(F_q^* \otimes F_q) \to 0$

(resp. the corresponding cohomology sequence in the category of

π -sheaves).

Now by the differential study of the scheme Q(E/P) as well as Q^{π} (cf. § 5, [6]), we deduce that the local ring of R (resp. R^{π}) at q is formally smooth (over $\mathcal{C}$) (cf. § 17, Chap. IV, [7]) i.e. if T is any affine scheme (over $\mathcal{C}$) and T_o a closed subscheme defined by an ideal I with $I^2 = 0$, then if $f_o : T_o \longrightarrow \mathrm{Spec}\ \mathcal{O}_{R,q}$ (resp. $\mathcal{O}_{R^{\pi},q}$) is a morphism it can be extended to a morphism $f : T \longrightarrow \mathrm{Spec}\ \mathcal{O}_{R,q}$ (resp. $\mathcal{O}_{R^{\pi},q}$), since $H^1(H_q^* \otimes F_q) = 0$ (resp. $H^1(X, \pi, H_q^* \otimes F_q) = 0$; for this case, one has to repeat the arguments of § 5, [6]).

It follows now that $\mathcal{O}_{R,q}$ (resp. $\mathcal{O}_{R^{\pi},q}$) is smoth (over $\mathcal{C}$) (cf. § 17, Chap. IV [7]), so that R (resp. R) is smooth.

This completes the proof of (iv).

(v) Again by the differential study of the scheme Q(resp. Q^{π}) referred to above (§ 5, [6]), $H^o(H_1^* \otimes F_q)$ (resp. $H^o(X, \pi, H_q^* \otimes F_q)$) can be identified canonically with the Zariski tangent space to R (resp. R^{τ}) at q.

Since R(resp. R) is smooth and irreducible, we have dim R(resp. dim R^{τ}) = dim $H^o(H_q^* \otimes F_q)$ (resp. $H^o(X, \pi, H_q^* \otimes F_q)$).

Now because of (A) above, we have

$$\dim R\ (\text{resp. } \dim R^{\tau}) = \dim H^o(E^* \otimes F_q - \chi(F_q^* \otimes F_q$$

$$= (\text{resp. } \dim H^o(X, \pi, E^* \otimes F_q) - \chi(F_q^* \otimes F_q)).$$

Now E^* is a trivial bundle, so that we have

$$H^o(E^* \otimes F_q) = H^o(E^*) \otimes H^o(F_q) = H^o(E^* \otimes E)\ .$$

It follows also that

$$H^o(X, \pi, E^* \otimes F_q) = H^o(X, \pi, E^* \otimes E).$$

Thus $H^o(E^* \otimes F_q)$ resp. $H^o(X, \pi, E^* \otimes F_q))$ identifies canonically with $H^o(E^* \otimes E)$ (resp. $H^o(X, \pi, E^* \otimes E))$ i.e. the space of endomorphisms (resp. π endomorphisms) of E.

The algebraic group of automorphisms of E (resp. π-automorphisms of E) is open in $H^o(E^* \otimes E)$ (resp. $H^o(X, \pi, E^* \otimes E)$).

This shows that

$$\dim R \text{ (resp. } \dim R^{\tau}) = \dim G \text{(resp. } \dim H)$$

$$- \chi(F_q^* \otimes F_q) \text{ (resp. } \chi(X, \pi, F_q^* \otimes F_q)).$$

Now $\deg(F_q^* \otimes F_q) = 0$ and $\operatorname{rk}(F_q^* \otimes F_q) = r^2$.

Therefore by the Riemann-Roch theorem, we get

$$- \chi(F_q^* \otimes F_q) = r^2 (g-1)$$

This proves that $\dim R = (r^2(g-1) + 1) + (\dim G - 1)$.

We have $\chi(X, \pi, F_q^* \otimes F_q) = \chi(Y, p_*^{\pi}(F_q^* \otimes F_q))$.

Again by the Riemann-Roch theorem, we have

$$- \chi(Y, p_*^{\pi}(F_q^* \otimes F_q)) = r^2(h-1) - \deg p_*^{\pi}(F_q^* \otimes F_q).$$

On the other hand, we have

$$- \deg p_*^{\pi}(F_q^* \otimes F_q) = \frac{1}{2} \sum_{\nu=1} e_{\nu}$$

as has been verified in the proof of Cor. 2, Th. 3, Chap. I.

Thus we have

$$\dim R^{\tau} = (r^2(h-1) + 1) + \frac{1}{2} \sum_{\nu=1} e_{\nu} + (\dim H - 1)$$

and the proposition is proved.

Remark 1. We have a little more than what is stated in (iii) of the above proposition.

We have in fact a neighbourhood T_o of t and a morphism $f : T_o \to R$ such the defining bundle on $X \times T_o$ of $\{V_t\}$ is the inverse image by Id f : $X \times T_o \to X \times R$ (resp. $X \times R^{\tau}$) of the family of vector bundles on $X \times R$ (resp. $X \times R^{\tau}$) defined by the restriction of F to $X \times R$ (resp. $X \times R^{\tau}$).

Corollary (of Proposition 6). (Local module for a vector bundle with trivial automorphism).

Let V_o be a vector bundle (resp. π-vector bundle) on X such that $\dim_C \operatorname{End} V_o$ (resp. $\dim_{\mathcal{C}} \operatorname{End}_{\pi} V_o$) = 1.

Then there is a holomorphic family $\{V_d\}$ parametrized by an analytic manifold D and a point D_o D such that

(1) $V_{d_o} \approx V_o$ and $V_{d_1} \not\approx V_{d_2}$ (i.e. not isomorphic) if $d_1 \neq d_2$, $d_2 \in D$.

(2) given a holomorphic family $\{W_t\}$ of vector bundles parametrized by an analityc space T and a point $t_o \in T$ such that $W_{t_o} \approx V_o$, there is an open neighbourhood T_o of T and a (unique) morphism $f : T_o \longrightarrow D$ such that the family $\{W_t\}_{t \in T_o}$ is the inverse image of the family $\{V_t\}_{t \in M}$ i.e. the defining bundle of $\{W_t\}_{t \in T_o}$ on $X \times T_o$ is the inverse image of the defining bundle of $\{V_d\}_{d \in D}$ by the analityc map $\operatorname{Id} \times f : X \times T_o \longrightarrow X \times M$.

We have in particular $V_d \approx W_t$ where $d = f(t)$, and

(3) $\dim D = \dim H^1 (V_o^* \otimes V_o$ (resp. $H^1(X, \pi, V_o^* \otimes V_o))$.

Proof. Choose an integer m such that (a) $H^1(V_o(m)) = 0$ and (b) $H^o(V_o(m))$ generates $V_o(m)$.

Let E be the trivial vector bundle $X \times H^o(V_o(m))$ (resp. the π-vector bundle associated to the canonical π-module $H^o(V_o(m))$).

Consider the scheme $Q(E/P)$ (resp. $Q^{\pi}(E/P)$), P being the Hilbert polynomial of $V_o(m)$.

Then there is a unique G orbit K in R (resp. H orbit K in R^{τ} for a suitable choice of m) such that

$F_q \approx V_o(m) \iff q \in K$ ($V_o(m)$ sheaf associated to $V_o(m)$).

We have $H \subset G$ and H contains the scalars matrices $\lambda \cdot \mathrm{Id}$, $\lambda \in \mathcal{C}$.

We note that the scalar matrices in G (resp. H) operate trivially on R (resp. R^{τ}).

Denote by PG (resp. PH), the quotient group of G (resp. H) modulo the group of scalar matrices.

Then the action of G (resp. H) on R (resp. R^{τ}) gives rise to a canonical action of PG (resp. PH) on R (resp. R^{τ}).

By our hypothesis if $q \in K$ then the isotropy group q for the action of PG (resp. PH) reduces to the identity element of PG (resp. PH).

Then there is a neighbourhood U of K (in the topology of analytic spaces) in the manifold R (resp. R^{τ}) such that PG (resp. PH) operates freely on U, the quotient $D = U/G$ (resp. U/H) exists as a manifold and in fact there is a <u>section</u> of R (resp. R^{τ}) over D.

Because of this section, we have a holomorphic family of vector bundles $\{V'_d\}$ $d \in D$ parametrized by D and a point $d_o \in D$ such that $V'_{d_o} \approx F_q \approx V_o(m)$ for $q \in K$ and $V'_d \approx F_q$, d being the image of $q \in U$ under the canonical morphism $U \longrightarrow D$.

Let $\{V_d\}$ $d \in D$ be the holomorphic family $V_d = V'_d(-m)$.

We have $V_{d_o} \approx V_o$.

Suppose now that $\{W_t\}_{t \in T}$ is a holomorphic family of vector bundles as in (2) of the Corollary; W being the defining bundle of $\{W_t\}$ on $X \times T$.

Then by the semi-continuity theorem, we can find an integer m and a neighbourhood T_o of t_o such that (a) $H^1(W_t(m)) = 0 \ \forall \ t \in T_o$ and (b) $H^o(W_t(m))$ generates $W_t(m) \ \forall \ t \in T_o$.

Then by the property (iii) of the above proposition, we have a morphism f_1 of a neighbourhood of t_o, which we can assume to be T_o, into R (resp. R^{τ}) such that if

$q = f(t)$, $F_q \approx V_t(m)$ or $F_q(-m) \approx V_t$.

We can also suppose that $f_1(T_o) \subset U$.

Let $f : T_o \longrightarrow D$ be the morphism obtained by composing f_1 with the canonical morphism $U \longrightarrow D = U/G$ (resp. U/H).

If $d = f(t)$, we defuce immediately that $W_t \approx V_d$, and in fact that the defining bundle of $\{W_t\}_{t \in T_o}$ is the inverse image by $Id \times f : X \times T_o \longrightarrow X \times D$ of the defining bundle of $\{V_d\}_{d \in D}$ (use also the remark on (iii) of the above proposition at the end of its proof).

Finally, we have $\dim D = \dim R - \dim PG$

$= $ (resp. $\dim R - \dim PH$).

But we have $\dim R = \dim PG + \dim H^1 (V_o^* \otimes V_o)$

$=$ (resp. $\dim PH + \dim H^1(X, \pi, V_o^* \otimes V_o)$).

This proves (3) of the corollary and the corollary is proved.

Remark 2. Let as usual S_r (resp. $S_{\pi, r}$) be the category of semi-stable (resp. π-semi-stable) vector bundles on X of rank r and degree zero (resp. π-degree zero $\Longleftrightarrow$ deg is zero) and S_{τ} the subcategory of $\mathbf{S}_{\pi, r}$ consisting of all π-semi-stable vector bundles V in $\mathbf{S}_{\pi, r}$ which are locally of a fixed type τ (cf. Remark 1, Proposition 2, Chapter I).

Then since $\mathbf{S}_r$ is bounded (cf. Prop. 3, § 1), we can find an integer m such that $H^0(V(m))$ generates V(m) and $H^1(V(m)) = 0 \ \forall \ V \in \mathbf{S}_r$.

Let P be the Hilbert polynomial of V(m), $V \in \mathbf{S}_r$.

Then every V(m) V(m), $V \in \mathbf{S}_r$ (resp. $V \in D_\tau$) can be considered as a quotient of the trivial vector bundle E of rank = = dim $H^0(V(m))$ (resp. π-quotient of a suitable π-bundle E such that the underlying vector bundle is trivial and of rank = dim $H^0(V(m))$, $V \in \mathbf{S}_r$).

Thus we can associate to each V(m), $V \in \mathbf{S}_r$ (resp. $V \in \mathbf{S}_\tau$) canonically a G-orbit (resp. H orbit) in R(E/P) (resp. R^π (E/P)).

This reduces the problem of classification (upto isomorphism) of semi-stable vector bundles (resp. π-semi-stable vector bundles) to a problem of constructing orbit spaces of the subset of R (resp. R^π or R^τ) corresponding to semi-stable (resp. π-semi-stable) vector bundles.

Proposition 7. Let V_0 be avector bundle (resp. π-vector bundle) associated to an irreducible unitary representation ρ of Γ_0 (resp. of Γ) of rank r.

Let $\{V_d\}_{d \in D}$ $(V_{d_0} \approx V_0)$ be the analytic family as in Cor. Prop. 6 (Chap. II) above.

Then there is a neighbourhood D_0 of d_0 such that $\forall \ d \in D_0$, V_d is isomorphic (resp. π-isomorphic) to a vector bundle (resp. π-vector bundle) associated to an irreducible unitary representation of Γ_0 (resp. Γ) (for the notation Γ_0, see at the beginning of Chap. II).

(ii) Let $\{V_t\}_{t \in T}$ be an analytic family of vector bundles (resp. π-vector bundles) on X.

Then the subset T_o of T of points t such that V_t is isomorphic (resp. π-isomorphic) to a vector bundle associated to an irreducible unitary representation of Γ_o (resp. Γ) is open.

Proof. Let S be the $\mathscr{C}$-analytic spaces of all representations of Γ_o in GL (r, $\mathscr{C}$) (resp. be the $\mathscr{C}$-analytic spaces of all representations χ of Γ into GL (r, $\mathscr{C}$) in X such that the π-vector bundle on X defined by χ is locally of type $\mathcal{T}$).

Let U denote the subset of S corresponding to unitary representations.

We see easily that there is an analytic family $\{W_s\}$ of vector bundles (resp. π-vector bundles) on X parametrized by S.

There is an $s_o \in U \subset S$ such that $V_{s_o} \approx V_o$.

By the universal property of $\{V_d\}$ of Cor. 1, Prop. 6 (Chap. II), we have a canonical analytic map of a neighbourhood of s_o in D and therefore its restriction to U defines a continuous map f of a neighbhourhood U_o of s_o into D. such that if $d = f(t)$, then $V_d \approx W_t$.

Since equivalence classes of unitary representation of Γ_o (resp. Γ) define isomorphic bundles, we can suppose that U_o is invariant under the canonical action of the unitary group K of rank r defining equivalent representations.

Thus f defines canonically a continuous map $g : U_o/K \longrightarrow D$ which is injective.

Because § 3, Chap. I, we can suppose that U_o again consists of points defining irreducible unitary representations of Γ_o (resp. Γ) and besides U_o/K is a manifold whose real dimension is equal to $\dim_{\mathcal{R}} H^1(\Gamma_o, \text{ad } \rho)$ (resp. $\dim_{\mathcal{R}} H^1(\Gamma, \text{ad } \rho)$).

But on the other hand, we have

$$\dim_{\mathcal{R}} H^1(\Gamma_o, \text{ad } \rho) = 2 \dim_{\mathcal{C}} H^1(V_o^* \otimes V))$$

(resp. $\dim_{\mathcal{R}} H^1(\Gamma, \text{ad } \rho) = 2 \dim_{\mathcal{C}} H^1(X, \pi, V^* \otimes V)$).

(cf. § 3, Th. 3, Chap. I and its corollaries). But we have $\dim D = \dim H^1(V_o^* \otimes V)$ (resp. $H^1(X, \pi, V^* \otimes V)$), dimension being as a complex manifold.

Therefore, as topological manifols, we have

$$\dim D = \dim U_o/K.$$

Since g is injective we conclude that g is a local homeomorphism by Browwer's theorem.

This proves (i) of Prop. 7.

Now (ii) is an immediate consequence of the local universal property of $\{V_d\}$, namely the property (2) of Cor. 1, Prop. 6 (Chap. II) and Prop. 7 is proved.

3. Some results from Mumford's geometric invariant theory.

In this section, we give a rapid survey of some results from Mumford's geometric invariant theory (Ch. 0, 1, 2, and § 4 Chap. 4, [9] ; and § 2, [18]) which we need.

In this section of Chapter, we do not conform to the notations mentioned at the beginning of this chapter. We keep still the convention that we consider only algebraic schemes defined over $\mathcal{C}$ and by a point of a scheme we maen a closed point unless otherwise stated.

C.S. Seshadri

Let X be an algebraic scheme on which an affine algebraic group G operates.

A morphism $\varphi : X \longrightarrow Y$ of algebraic schemes is said to be a good quotient (of X modulo G) if it has the following properties:

(i) φ is G-invariant i.e. the followinf diagram

$$\begin{array}{ccc} X \times G & \xrightarrow{\chi} & X \\ {\scriptstyle pr_X}\downarrow & & \downarrow{\scriptstyle \varphi} \\ X & \xrightarrow{\varphi} & Y \end{array}$$

is commutative, where $\chi : X \quad G \longrightarrow X$ is the morphism by which the action of G on X is defined.

(ii) φ is a surjective affine morphism

(iii) $(\varphi_*(0_X))^G = 0_Y$

(iv) if X_1, X_2 are closed G-invariant subsets of X such that $X_1 \cap X_2$ is empty, then $\varphi(X_1)$, $\varphi(X_2)$ are closed and $\varphi(X_1) \cap \varphi(X_2)$ is empty.

We say that $\varphi : X \longrightarrow Y$ is a good affine quotient if φ is a good quotient and Y is affine (this imples that X is also affine).

The first three conditions are equivalent to the following : φ is surjective and for every affine open subset U of Y , $\varphi^{-1}(U)$ is affine and G-invariant and the coordinate ring of U can be identified with the G-invariant suring of $\varphi^{-1}(U)$.

Some properties of good quotient are collected together in the following.

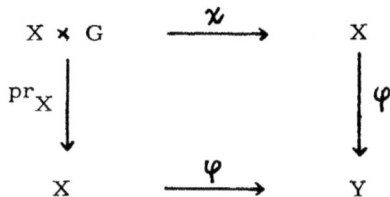

Proposition 8. (i) The property of being a good quotient is local with respect to the base scheme i.e. $\varphi : X \longrightarrow Y$ is a good quotient if and only if there is an open covering U_i of Y such that every $V_i = \varphi^{-1}(U_i)$ is G-invariant and the induced morphism $\varphi_i : V_i \longrightarrow U_i$ is a good quotient.

(ii) a good quotient is also a cateorical quotient i.e. if $\varphi : X \longrightarrow Y$ is a good quotient, then for every G-invariant morphism $\psi : X \longrightarrow Z$, there is a unique morphism $\nu : Y \longrightarrow Z$ such that $\nu \circ \varphi = \psi$.

(iii) transitivity properties. Let X be an algebraic scheme on which an affine algebraic group G operates.

Let N be a normal closed subgroup of G and H the affine algebraic group G/N.

Suppose that $\varphi_1 : X \longrightarrow Y$ is a good quotient (resp. good affine quotient) of X modulo N.

Then we have the following :

(a) the action of G goes down into an action of H on Y

(b) if $\varphi_2 : Y \longrightarrow Z$ is a good quotient (resp. good affine quotient) of Y modulo H, then $\varphi_2 \circ \varphi_1 : X \longrightarrow Z$ is a good quotient (resp. good affine quotient) of X modulo G.

$$\begin{array}{ccc} X & \xrightarrow{\text{good}} & Y \\ & \searrow \quad \swarrow & \\ & Z & \end{array}$$

(c) if $\varphi : X \longrightarrow Z$ is a good quotient (resp. good affine quotient) of X modulo G, there is a canonical morphism $\varphi_2 : Y \longrightarrow Z$ such that $\varphi = \varphi_2 \circ \varphi_1$ and φ_2 is a good quotient (resp. good affine quotient) of Y modulo H.

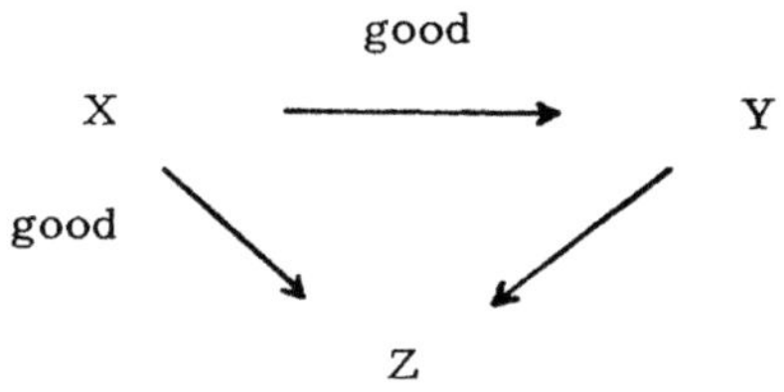

(d) if $\varphi : X \longrightarrow Y$ is a good quotient (modulo G), Z a <u>normal</u> algebraic variety on which G operates and $j : Z \longrightarrow X$ a proper injective G-morphism (in particular a closed immersion which is a G-morphism), then Z has a good quotient modulo G; in fact it can be identified with the normalization of the reduced subvariety $(\varphi \circ j)(Z)$ in a suitable finite extension of the field of rational functions of $(\varphi \circ j)(Z)$.

The proof of this proposition is quite easy and we leave it as an exercise.

The basic existence theorem on good quotient is the following :

<u>Theorem 1.</u> Let $X = \text{Spec } A$ be an affine algebraic scheme on which a <u>reductive</u> affine algebraic group G operates (note that we have supposed the ground field to be $\mathcal{C}$).

Let $Y = \text{Spec } A^G$ (A^G - G invariant subring of A) and $\varphi : X \longrightarrow Y$ the canonical morphism induced by $A^G \subset A$.

Then Y is an affine algebraic scheme and $\varphi : X \longrightarrow Y$ is a good affine quotient.

<u>Outline of proof.</u> It is a classical fact that A^G is finitely generated over $\mathcal{C}$.

Therefore Y is an affine <u>algebraic</u> scheme.

Let V be a finite dimensional vector space over C and be a G-module through a homomorphism $\rho : G \longrightarrow GL(V)$ of algebraic groups.

Then G being reductive, V is a semi-simple G-module and we have a canonical linear projection $V \longrightarrow V^G$ which is functorial in V, called the Reynold's operator.

Now given $f \in A$, it can be imbedded in a finite dimensional G-submodule V of A (for the G-module structure on V induced by the right or left regular representation).

Because of this, we get a canonical linear projection $p : A \longrightarrow A^G$.

Suppose now that X_1, X_2 are two closed G-invariant subsets of X such that $X_1 \cap X_2$ is empty.

Then there exists $f \in A$ such that f is 0 on X_1 and 1 on X_2.

Let $g = p(f)$.

Then $g \in A^G$ and we see again that g is 0 on X_1 and 1 on X_2 (this results from the functoriality of the Reynold's operator).

Thus we can <u>separate</u> X_1 and X_2 by a G-invariant function g on X.

This is the crucial property and this implies easily the property (iv) in the definition of good quotients and the proposition follows.

For more details § 2, Chap. I, [9] .

Let X be a closed subscheme of the projective space ${}_{n}$ of dimension n.

An action of an affine algebraic group G on X is said to be <u>linear</u> if it comes from a rational representation of G in the affine scheme $\mathcal{A}_{n+1}$ of dimension $(n+1)$. This definition means that we have an action of G on $\mathcal{A}_{n+1} = \text{Spec } \mathcal{C}\,[X_1, \ldots, X_{n+1}]$ given by a rational representation of G on A_{n+1} and that if $\mathfrak{U}$ is the graded ideal of $\mathcal{C}[X_1,\ldots,X_{n+1}]$ defining X, then $\mathfrak{U}$ is G-invariant.

We have $X = \mathrm{Proj.}\ R$, where $R = \mathcal{C}[X_1,\ldots,X_{n+1}]$.

We denote by $\hat{X}$ the cone over X (X = Spec R) and by (0) - the vertex of the cone $\hat{X}$.

The action of $\mathcal{G}$ on $\hat{X}$ lifts to an action of G on X and this action and the canonical action of the multiplicative group $\mathcal{G}_m$ on $\hat{X}$ (by homothecies) commute.

We observe that the canonical morphism $p : \hat{X} - (0) \longrightarrow X$ is a principal fibre space with structure group $\mathcal{G}_m$ and that p is a good quotient (modulo $_m$).

Suppose then that X is a closed subscheme of $\mathcal{P}_n$ and that we have a linear action of an affine algebraic group G on X.

A point $x \in X$ is said to be semi-stable if for some $\hat{x} \in \hat{X} - (0)$ over x, the closure (in $\hat{X}$) of the G-orbit through $\hat{x}$ does not pass through (0).

A point $x \in X$ is said to be stable (to be more precise, properly stable) if for some $\hat{x} \in \hat{X} - (0)$ over x, the orbit morphism $\Psi_{\hat{x}} : G \longrightarrow \hat{X}$ defined by $g \longmapsto \hat{x} \circ g$ is proper.

Since the actions of $\mathcal{G}$ and $\mathcal{G}_m$ on $\hat{X}$ commute, we see easily that if $x \in X$ is semi-stable, then for every $\hat{x} \in \hat{X} - (0)$ over x, the closure (in $\hat{X}$) of the G-orbit through $\hat{x}$ does not pass through (0).

A similar property holds for the stable points of X.

We denote by X^{ss} (resp. X^s) the set of semi-stable (resp. stable) points of X.

With these definitions and notations, we have the following.

Theorem 2. Let X be a closed subscheme of $\mathcal{P}_n$ defined by a grated ideal $\mathfrak{U}$ of $\mathcal{C}[X_1,\ldots,X_{n+1}]$ so that X = Proj. R,

$R = \mathscr{C}[X_1, \ldots, X_{n+1}]$.

Let there be given a linear action of a <u>reductive</u> affine algebraic group G on X.

Let $Y = \text{Proj. } R^G$ and $\varphi : X \longrightarrow Y$ the canonical <u>rational</u> morphism defined by the inclusion $R^G \subset R$.

Then we have

(i) $x \in X^{ss}$ if and onli if there is a homogeneous G-invariant element $f \in R$ (R_+ being the subring of R generated by homogeneous elements of degree ≥ 1) such that $f(x) \neq 0$.

Note that X_f is a G-invariant affine open subset of X and $x \in X_f \subset X^{ss}$ so that we have in particular that X^{ss} is open in X and for every $x \in X^{ss}$, there is a G-invariant affine open subset containing x and contained in X^{ss}.

Further (i) implies that φ is defined at $x \in X^{ss}$.

(ii) $\varphi : X^{ss} \longrightarrow Y$ is a good quotient and Y is a projective algebraic scheme and

(iii) X^s is a φ-saturated open subset i.e. there exists an open subset Y^s of Y such that $X^s = \varphi^{-1}(Y^s)$ and $\varphi : X^s \longrightarrow Y^s$ is a <u>geometric quotient</u> i.e. distinct orbits (under G) of X^s go into distinct points of Y^s.

<u>Outline of proof.</u> Let $\hat{Y} = \text{Spec } R^G$ and $\hat{\varphi} : \hat{X} \longrightarrow \hat{Y}$ the canonical morphism induced by the inclusion $R^G \subset R$.

Then by Theorem 1, $\hat{\varphi} : \hat{X} \longrightarrow \hat{Y}$ is a good affine quotient.

From this it follows easily that $x \in X^{ss}$ if and only if there is an $f \in R^G$ such that $f(0)) = 0$ and $f(\hat{x}) \neq 0$, where $\hat{x}$ is some point in $\hat{X} - (0)$ over x.

Now the homogeneous components of f are also G-invariant

and $f((0)) = 0$ implies that there is a homogeneous components f_d of f such that f_d is in R_+ and $f_d(x) \neq 0$.

Thus we see that $x \in X^{ss}$ if and only if there is a homogeneous f in R_+ such that $f(x) \neq 0$.

Now the canonical morphism $\hat{\varphi}_f : \hat{X}_f \longrightarrow \hat{Y}_f$ induced by φ is a good affine quotient by Theorem 1 (Chap. II).

Since R^G is finitely generated over $\mathscr{C}$ $Y = \mathrm{Prof}\ R^G$ is a projective algebraic scheme.

We see easily that we have a canonical morphism $\varphi_f : X_f \longrightarrow Y_f$ induced by the inclusion $(R_f^G)^o \subset (R_f)^o$ where $(R_f)^o$ (resp. $(R_f^G)^o$) indicates the homogeneous elements of degree 0 in the localization R_f of R (resp. R_f^G of R^G) with respect to the multiplicative closed subset of R (resp. R^G) formed by powers of f.

By the local nature of good quotients (cf. (i) of Prop. 8 Chap. II), it suffices to prove that $\varphi_f : X_f \longrightarrow Y_f$ is a good quotient.

But now we observe that the coordinate ring of the affine scheme Y_f is $(R_f^G)^o$ which is precisely $(R_f^o)^G$ i.e. it is the G-invariant subring of R_f^o which is the coordinate ring of the affine scheme X_f.

Therefore $\varphi_f : X_f \longrightarrow Y_f$ is a good quotient by Theorem 1 (Chap. II).

This proves the assertions (i) and (ii).

Let $\hat{X}^s$ denote the set of points $\hat{x} \in \hat{X}$ such that the orbit morphism $\Psi_{\hat{x}} : G \longrightarrow \hat{X}$ is proper.

Then $\hat{X}^s$ is $p^{-1}(X^s)$, where p is the canonical morphism $\hat{X} - (0) \longrightarrow X$.

Let U be the subset of $\hat{X}$ consisting of the points $\hat{x} \in \hat{X}$ such that $\dim G = \dim \Psi_{\hat{x}}(G)$.

By an easy application of the dipmension theorem, we see that U is open.

Further U is obviously G-invariant.

Let $W = \hat{X} - U$.

Then $C = \hat{\varphi}(W)$ is a closed subset of Y because $\hat{\varphi} : \hat{X} \longrightarrow \hat{Y}$ is a good quotient.

It is easily seen that $\hat{X}^s = \hat{\varphi}^{-1}(\hat{Y} - C)$.

This implies that $\hat{X}^s$ is a φ-saturated open sunset of $\hat{X}$.

From this it follows easily that X^s is a φ-saturated open subset of X.

Let now $x_1, x_2 \in X^s$ such that the G-orbits $0_1, 0_2$ (in X) through x_1, x_2 respectively have the property $0_1 \cap 0_2 \cap X^s$ is empty.

Since X^s is φ-saturated, it follows that $0_1 \cap 0_2 \cap X^{ss}$ is empty.

Therefore, for every G-invariant $f \in R_+$, we have $0_1 \cap 0_2 \cap X_f$ is empty since $X_f \subset X^{ss}$.

From the fact that $\varphi_f : X_f \longrightarrow Y_f$ is a good quotient, we deduce easily that $\varphi(x_1) \neq \varphi(x_2)$ (we observe that X_f is also φ-saturated).

This proves (iii) and thereby the theorem.

Let $H_{p,r}(E)$ denote the Grassmannian of r-dimensional quotient linear spaces of a p-dimensional vector space E(over $\mathscr{C}$).

We have a canonical immersion of $H_{p,r}(E)$ into the projective space associated to $\overset{p-r}{\bigwedge} E$ and if $X = H^N_{p,r}(E)$ denotes the

N-fold product of $H_{p,r}(E)$, we have a canonical projective imbedding of X, namely the Segre imbedding of X associated to the canonical projective imbedding of $H_{p,r}$4E).

There is a natural action of GL(E) on $H_{p,r}(E)$ and this induces a natural action (diagonal action) of GL(E) on $H^N_{p,r}(E)$.

The restriction of this action to the subgroup G = SL(E) of GL(E) is a <u>linear</u> action with respect to the canonical imbedding of X;

We denote by X^{ss}(resp. X^s) the set of semi-stable (resp. stable) points of X for the action of G with respect to the canonical projective imbedding of X.

With the above notation, we have the following important computational result of Mumford :

<u>Theorem 3.</u> Let $X = H^N_{p,r}(E)$ and X^s, X^{ss} be as above. Then for $x \in X$, $x = \{E_i\}_{1 \leq i \leq N}$, E_i a quotient linear space of dimension r of E, $x \in X^{ss}$ (resp. X^s) if and only if for every linear subspace (resp. proper linear subspace) F of E, if F_i denotes the canonical image of F in E_i, we have

$$\frac{\frac{1}{N}\sum_{i=1}^{N} \dim F_i}{r} \geq \frac{\dim F}{p} \quad (\text{resp. } >).$$

<u>Outline of proof.</u> (a) Let us call a one parameter subgroup of an algebraic group G (abbreviated 1 - PS of G) a rational homomorphism of $\mathcal{G}_m$ into G.

Let there be given a linear action of a reductive algebraic group G on a projective scheme $X \subset \mathcal{P}_n$.

Then a basic result is that $x \in X^{ss}$ (resp. X^s) if and only if it is so with respect to the restriction of the action of G to every 1 - PS of G (cf. Page 53, Chap. 2, [9]).

This fact can be expressed in a quantitative manner as follows:

Suppose there is an action of $\mathcal{G}_m$ on the projective space $\mathcal{P}_n$ induced by a linear action of $\mathcal{G}_m$ on the affine space $\mathcal{A}_{n+1}$.

With respect to a suitable coordinate system in $\mathcal{A}_{n+1}$, this action is given by

$x = (x_0, \ldots, x_n) \mapsto (\alpha^{r_0} x_0, \ldots, \alpha^{r_n} x_n)$, where α is the canonical coordinate of $\mathcal{G}_m$.

One defines for $x \in \mathcal{P}_n$, $\mu(x) = \max \{-r_i \mid i$ such that $x_i^* \neq 0\}$ where $x^* = (x_i^*)$ is a point of $\mathcal{A}_{n+1}$ over x P_n.

In this manner, for the action of G on X, we obtain an integer $\mu(x, \lambda)$ for every $x \in X$ and every 1 - PS λ of G.

We note that if y is the specialization of $\alpha \cdot x$ as $\alpha \to 0$, $\mu(x, \lambda) = \mu(y, \lambda)$.

The above result can be expressed as follows :

$x \in X^{ss}$ (resp. X^s) $\Leftrightarrow$ $\begin{cases} \mu(x, \lambda) \geqslant 0 \quad \text{(resp. } > 0) \\ \text{or } \mu(y, \lambda) \geqslant 0 \quad \text{(resp. } > 0), \ y \text{ being the} \\ \text{specialization of } \alpha \cdot x \text{ as } \alpha \to 0. \end{cases}$

(cf. § 1, Chap. 2, [9]).

The proof of (a) is not difficult.

By the definition of stable and semi-stable points of $X \subset \mathcal{P}_n$, we are reduced to proving the following : we are given a linear action of G (assumed reductive, affine) on the affine space $\mathcal{A}_{n+1}$.

Then if $x \in \mathcal{A}_{n+1}$ such that the orbit morphism $\Psi_{x,G} : G \longrightarrow \mathcal{A}_{n+1}$ with respect to G is <u>not proper</u> (resp. the closure of $\Psi_{x,G}(G)$ does not pass through the origin in $\mathcal{A}_{n+1}$) then there is a 1 - PS λ of G such that the orbit morphism $\Psi_{x,\lambda} : \mathcal{G}_m \longrightarrow A_{n+1}$ with respect to λ is also <u>not proper</u> (resp. the closure of $\Psi_{x,\lambda}(\mathcal{G}_m)$ does not pass through the origin in $\mathcal{A}_{n+1}$).

The proof of this is given on pages 53-54 of §1, Chap. 2 [9] .

(Note, incidentally that for the case when G = GL(r) or SL(r), a theorem of Iwahori which is used in [9] in the proof, is quite easy).

(b) Let $X = H^N_{p,r}(E)$. If $x \in X$, let us write

$$x = (L', \dots, L^N)$$

where L^i, $1 \leq i \leq N$, denotes a k-dimensional hyperplane of the projective space P(E) associated to E (in the usual sense).

If we take a 1-PS λ of SL(E) in the diagonal form, say $\lambda(\alpha) = (\alpha^{r_i} \delta_y)$, $0 \leq i, j \leq p-1$, and $r_o > r_1 > \dots > r_{p-1}$, the specialization y of $\alpha \cdot x$ can be computed explicitly and thereby a formula for $\mu(x, \lambda) = \mu(y, \lambda)$ is obtained.

This formula holds even when $r_o \geq r_1 \geq \dots \geq r_{p-1}$ and being a linear function of r_i $\mu(x, \lambda)$ is positive for every 1-PS λ in the diagonal torus T of SL(E) if and only if it is so for the extreme sets of r_i, namely when

$(p-1-q) = r_o = \dots = r_q$; $r_{q+1} = \dots = r_{p-1} = -(q+1)$.

Writing down this condition and writing it in an SL(E) invariant form, we get the theorem (for the details see § 4, Chap. 4, [9]).

This theorem is quite non trivial and represents one of the significant computations made in [9]).

4. Stable bundles and stable points in $H^N_{p,r}(E)$.

We follow now the notations and conventions made at the beginning of this chapter.

Consider the category $\mathcal{S}_r$ of semi-stable vector bundles on X of rank r and degree 0.

Then if m is an integer which is sufficiently large, we have $H^1(V(m)) = 0$ and $H^o(V(m))$ generates $V(m)$, for every $V \in \mathcal{S}_r$ (cf. Prop. 3, § 1).

Fix such an integer m.

Let P be the Hilbert polynomial of $V(m)$, $V \in \mathcal{S}_r$ and E the trivial vector bundle on X of rank = dim $H^o(V(m))$.

Let

$Q = Q(E/P)$, $R = R(E/P)$ and $R_1 = R(E/P)$ be as in § 2. Let R^{ss} (resp. R^s) denote the subset of $q \in R$ such that the vector bundle associated to $\mathfrak{F}_q$ is semi-stable (resp. stable).

Let us denote by F_q the vector bundle associated to the coherent sheaf $\mathfrak{F}_q$ and by F the vector bundle on $X \times R_1$ associated to the restriction of the defining sheaf $\mathfrak{F}$ of $\{F_q\}$ to $X \times R_1$.

Let $\mathfrak{n}$ denote an ordered set of N points $P_1, \dots, P_N$ on the curve X.

Let $\chi_i : R_1 \longrightarrow H_{p,r}(E)$ be the morphism into the Grassmannian of r dimensional quotient linear spaces of the p-dimensional vector space $H^o(E)$ (we set $p = \dim H^o(E)$) which assigns to $q \in R_1$ the fibre at P_i of the vector bundle F_q considered canonically as a quotient linear space of $H^o(E)$ (if we conform to the notations of 3, we should write $H_{p,r}(H^o(E))$ instead of $H_{p,r}(E)$.

We use this notation for simplicity).

Let

$$\chi : R_1 \longrightarrow H^N_{p,r}(E)$$

be the morphism defined by $\chi(q) = \{\chi_i(q)\}$., $1 \leq i \leq N$.

Let $G = GL(E0$ be the group of automorphism of the vector bundle E.

Then we have $G = GL(H^o(E))$ and we see that χ commutes with thecanonical actions of G on R_1 and $H^N_{p,r}(E)$ respectively (for the latter it is the diagonal action, cf. § 3).

We shall now extend the above morphism $\chi : R_1 \longrightarrow H^N_{p,r}(E)$ to a <u>multivalued set mapping</u> of $Q = Q(E/P)$ into $H^N_{p,r}(E)$ and we shall denote this extension by

$$\Phi = \{\Phi\}_i \quad , \quad \Phi_i : Q_1 \longrightarrow H_{p,r}(E), \quad 1 \leq i \leq N.$$

(if one prefers Φ is a subset of $Q_1 \times H^N_{p,r}(E)$ and Φ_i is a subset of $Q_1 \times H_{p,r}(E)$, $1 \leq i \leq N$).

Suppose now that for $q \in Q$ $\mathcal{F}_q$ is <u>not</u> locally free i.e. $q \notin R_1$.

Then we have $\mathcal{F}_q = V_q \oplus T_q$, where V_q is locally free and T_q is a torsion sheaf (because X is a smooth curve).

Suppose that $P_i \notin$ Supp. T_q (support of T_q).

We then define $\Phi_i(q) \in H_{p,r}(E)$ as the fibre of the vector V_q associated $\mathcal{V}_q$ at P_i considered canonically as a quotient linear space of dimension r of E.

Suppose that $P_i \in$ Supp. T_q; we then define $\Phi_i(q)$ to be any point of $H_{p,r}(E)$.

We thus obtain a multivalued (set) mapping $\Phi_i : Q \to H_{p,r}(E)$.

It is easy to see that Φ_i is a morphism in a neighbourhood of $q \in Q$ if and only if $P_i \in$ Supp. T_q(for by Lemma 4 of § 2 (Chap. II), we see that the defining sheaf $\mathcal{F}$ of $\{F_q\}$ is locally free in a neighbourhood of $(P_i \times q)$ in $X \times Q$ which implies easily that Φ_i is a morphism in a neighbourhood of q).

We see immediately that the graph of Φ_i in $Q \times H_{p,r}(E)$ is closed and contains thec closure of the graph of $\chi_i : \mathbf{R}_1 \longrightarrow H_{p,r}(E)$.

From this, it follows easily that graph of Φ in $Q \times H^N_{p,r}(E)$ contains the closure of the graph of $\chi : R_1 \longrightarrow H^N_{p,r}(E)$.

Then with these notations we have the following basic fact which connect stable (resp. semi-stable) bundles with the stable (resp. semi-stable points) in $H^N_{p,r}(E)$ for the canonical action of G.

Proposition 9. If m and N (N = cardinality of the set $\mathfrak{n}$ of points $P_1, \ldots P_N$) are sufficiently large, then for $q \in Q = Q(E/P)$, $\Phi(q)$ is a semi-stable (resp. stable) point of $H^N_{p,r}(E)$ for the canonical action of G if and only if $q \in R^{ss}$ (resp R^s) i.e. the vector bundle F_q is semi-stable (resp. stable).

The proof of this proposition though not difficult requires some careful analysis.

To prove the proposition is to prove equivalently the following two assertions ; namely

(A)

$$q \in R^{ss} \Longrightarrow \Phi(q) = \chi(q) \in H^N_{p,r}(E)^{ss}$$

$q \in R^{ss}$ then $\Phi(q)$ is in $H^N_{p,r}(E)^s$ if and only if $q \in R^s$

and

(B) $\Phi(q) \in H^N_{p,r}(E)^{ss} \Longrightarrow q \in R^{ss}$

The proof of (A) is to be found in § 7 especially Prop. 7.2, Prop. 7.3 and Th. 7.1 [17] .

The proof of (B) is given in § 3, Lemma 2, - [18] and is more delicate than that of (A).

Here, we outline only a proof only of (A) but not of (B).

Outline of proof of (A).

We clain for sufficiently large m, we have the following:

(a) if $\mathcal{G}$ is any sub-bundle of V, $V \in S_r$ such that deg $G = 0$, then $H^1(\mathcal{G}(m)) = 0$ and $H^o(G(m))$ generates $\mathcal{G}(m)$.

Further, we have

$$\frac{\dim H^o(\mathcal{G}(m))}{\text{rk } \mathcal{G}(m)} = \frac{\dim H^o(V(m))}{\text{rk } V(m)}$$

(b) let $\mathcal{G}$ be any sub-bundle of V, $V \in S_r$ such that deg $\mathcal{G} < 0$ and $H^o(G(m)$ generates G(m) generically (i.e. there is at least one point $x \in X$ such that $H^o(G(m))$ generates the fibre of G(m) at x).

For the proof of (a), we observe that if G satisfies the hypothesis of (a), then G is semi-stable i.e. $G \in S_k$, $1 \leq k \leq r$.

Now for sufficiently large m, we have $H^0(V(m))$ generates $V(m)$ and $H^1(V(m)) = 0$ for every $V \in \mathbf{S}_k$, $1 \leq k$.

The equality

$$\frac{\dim H^0(\mathcal{G}(m))}{\mathrm{rk}\ \mathcal{G}(m)} = \frac{\dim H^0(V(m))}{\mathrm{rk}\ V(m)}$$

is an immediate consequence of the Riemann-Roch theorem.

For the proof of (b), we proceed as follows.

We have the following simple

Lemma 5. Let V ba a vector bundle on X such that $H^0(V)$ generates V generically.

Then we have

$$\dim H^0(V) \leq \deg V + \mathrm{rk}\ V.$$

The proof of this lemma is quite simple (cf. Lemma 7. 2 [17]).

If $\mathrm{rk}\ V = 1$, the hypothesis imples that V can be defined by an effective divisor D and the above inequality is obtained by induction on $\deg D$.

Then the general case is obtained by induction on $\mathrm{rk}\ V$.

To continue the proof of (b), we observe that if θ is an integer and $\mathcal{B}_1$ is the category formed of vector bundles $\mathcal{G}$ on X such that $\mathcal{G}$ is a sub-bundle of some $V \in \mathbf{S}_r$ and $\deg \mathcal{G} \geq \theta$, then $\mathcal{B}_1$ is bounded.

For proving this, we note that since $\mathbf{S}_r$ is bounded, there is an integer e such that whenever W is an indecomposable vector bundle on X such that $\deg W \geq e$ and $\mathrm{rk}\ W \leq r$, the only homomorphism of W into any V, $V \in S_r$ is the zero one (cf. Prop. 11.1, [12]).

This implies that the degrees of the indecomposable components of any $\mathcal{G} \in \mathcal{B}_1$ are both bounded above and below.

Then by a theorem of Atiyah (cf. p. 426, Th. 3, [1]) it follows that $\mathcal{B}_1$ is bounded.

Fix now the integer θ so that whenever $\mathcal{G}$ is a sub-bundle of some $V \in S_r$ and satisfies the condition deg $\mathcal{G} < \theta$, we have

$$\frac{\deg \mathcal{G}}{\operatorname{rk} \mathcal{G}} < -g .$$

Let $\mathcal{G}$ be a sub-bundle of some $V \in S_r$ such that $H^o(\mathcal{G}(m))$ generates $\mathcal{G}(m)$ generically and deg $\mathcal{G} < \theta$.

Then by Lemma 5, we have

$$\dim H^o(\mathcal{G}(m)) \leq \deg \mathcal{G}(m) + \operatorname{rk}(\mathcal{G}(m))$$

so that

$$\frac{\dim H^o(\mathcal{G}(m))}{\operatorname{rk} \mathcal{G}(m)} \leq \frac{\deg \mathcal{G}(m)}{\operatorname{rk} \mathcal{G}(m)} + 1 = \frac{\deg \mathcal{G}}{\operatorname{rk} \mathcal{G}} + 1 + m \deg L$$

where L is the line bundle defined by $\mathcal{O}_X(1)$.

On the other hand since for sufficiently large m, we have $H^1(V(m)) = 0$ for every $V \in S_r$, by applying the Riemann-Roch theorem we get

$$\frac{\dim H^{o}(V(m))}{rk\ V(m)} = -g + 1 + m \deg L, \quad m \text{ sufficiently large}.$$

Since we have $-g > \deg \mathcal{G}/rk\ \mathcal{G}$, (b) is proved in this case.

Suppose now $\deg \mathcal{G} \geq \theta$ and $\mathcal{G}$ is a sub-bundle of some $V \in S_r$ i.e. $\mathcal{G} \in \mathcal{B}_1$.

Then for sufficiently large m, we have also $H^1(\mathcal{G}(m)) = 0$.

Then by the Riemann-Roch theorem we have

$$\frac{\dim H^{o}(\mathcal{G}(m))}{rk\ \mathcal{G}(m)} = \frac{\deg \mathcal{G}}{rk\ \mathcal{G}} - g + 1 + m \deg L, \text{ for } m \text{ sufficiently large.}$$

We have $\deg \mathcal{G} < 0$ and this implies immediately that

$$\frac{\dim H^{o}(\mathcal{G}(m))}{rk\ \mathcal{G}(m)} < \frac{\dim H^{o}(V(m))}{rk\ V(m)}$$

i. e. (b) is proved.

Choose now an integer m such that the above properties (a) and (b) above hold.

Let $V \in S_r$.

Then if L is a proper linear subspace of $H^{o}(V(m))$, we set

$$\rho(L) = \frac{\frac{1}{N}\sum_{i=1}^{N} \dim L_i}{\dim L} - \frac{r}{p}$$

where $p = \dim H^o(V(m)) = rk\ E$ and L_i denotes the canonical image of L in the fibre of $V(m)$ at P_i $(P_1, \ldots, P_N$ the ordered set $\mathfrak{n}$ of points on X).

To prove (A), we have to show that $\rho(L) \geq 0$; further if V is semi-stable and not stable there is an L such that $\rho(L) = 0$ and that if V is stable $\rho(L) > 0$.

Let $\mathcal{G}$ be the unique sub-bundle of $V(m)$ generated by L ; set

$$\rho_1(L) = \frac{rk\ \mathcal{G}}{\dim L} - \frac{r}{p}$$

$$\rho_2(L) = \frac{rk\ \mathcal{G}}{\dim H^o(\mathcal{G})} - \frac{r}{p} \quad .$$

Suppose now that

(i) $$\frac{\deg\ \mathcal{G}}{rk\ \mathcal{G}} = \frac{\deg\ V(m)}{rk\ V(m)}$$

and that $L = H^o(G)$.

In this case V is semi-stable but not stable.

Then we have

$$\rho(L) = \rho_1(L)$$

further $\rho_1(L) = 0$.

This implies that $\rho(L) = 0$.

Suppose that

(ii) $\dfrac{\deg \mathcal{G}}{\operatorname{rk} \mathcal{G}} = \dfrac{\deg V(m)}{\operatorname{rk} V(m)}$ and $L \neq H^{o}(G)$.

Then $\rho_1(L) > \rho_2(L)$.
But $\rho_2(L) = 0$ by (a).
Therefore we have $\rho_1(L)$ 0. Suppose that

(iii) $\dfrac{\deg \mathcal{G}}{\operatorname{rk} \mathcal{G}} < \dfrac{\deg V(m)}{\operatorname{rk} V(m)}$.

Then $\rho_1(L) \geqslant \rho_2(L)$ and $\rho_2(L) > 0$ by (b) above.
Therefore again in this case $\rho_1(L) > 0$.
Thus when (i) does not hold, we have $\rho_1(L) > 0$.
Now we have

$$\rho_1(L) - \rho(L) = \frac{\frac{1}{N} \sum_{i=1}^{N} (\operatorname{rk} \mathcal{G} - \dim L_i)}{\dim L}$$

Let λ be the number of distinct points $x \in X$ such that L does not generate the fibre of G at x.

Then we have

$$(*) \qquad \frac{\lambda . \operatorname{rk}(\mathcal{G})}{N . \dim L} \geqslant (\rho_1(L) - \rho(L)) \geqslant 0.$$

Now we have the following

Lemma 6. Let W be a vector bundle on X and M a subspace of $H^o(W)$ such that M generates W generically.

Let μ be the number of distinct points $x \in X$ such that M does not generate the fibre of W at x.

Then we have

$$\mu \leq \deg W .$$

The proof of the lemma is quite easy (cf. Lemma 7.1, [17]) and we do not give it here.

Now by Lemma 6, (*) implies

$$\frac{\deg \mathcal{G} . \operatorname{rk}(\mathcal{G})}{N. \dim L} \geq (\rho_1(L) - \rho(L)) \geq 0 .$$

We have

$$\frac{\deg \mathcal{G}}{\operatorname{rk} \mathcal{G}} \leq \frac{\deg V(m)}{\operatorname{rk} V(m)}$$

Therefore, we have

$$\frac{\deg V(m) . \operatorname{rk}(\mathcal{G})^2}{N. \operatorname{rk} V(m). \dim L} \geq (\rho_1(L) - \rho(L)) \geq 0 .$$

Since $\dim L \geq 1$, we have in fact

$$(**) \qquad \frac{\deg V(m) . \operatorname{rk}(\mathcal{G})^2}{N. \operatorname{rk} V(m)} \geq (\rho_1(L) - \rho(L)) \geq 0.$$

Now if N is sufficiently large, we conclude we have in fact $\rho(L) > 0$ in cases (ii) and (iii) since we had already $\rho_1(L) > 0$.

This proves the assertion (A).

Corollary 1 (of Prop. 9). Let $\{V_t\}_{t \in T}$ be an algebraic or analytic family of vector bundles on X of degree 0.

Then the subset T^{ss} (resp. T^s) consisting of points $t \in T$ such that V_t is semi-stable is open in T.

In particular R^{ss} (resp. R^s) is an open subset of R (notations as in the above proposition).

Proof. Consider the multivalued se mapping $\Phi : Q \to H^N_{p,r}(E)$ for sufficiently large m and N as in Prop. 9.

Let $\chi : R \longrightarrow H^N_{p,r}(E)$ denote the porphism induced by Φ.

Then

$$R^{ss} \text{ (resp. } R^s) = \chi^{-1} (H^N_{p,r}(E)^{ss}) \text{ (resp. } \chi^{-1}(H^N_{p,r}(E))^s).$$

Since $H^N_{p,r}(E)^{ss}$ (resp. $H^N_{p,r}(E)^s$) is open (cf. 3, Th. 2, Chap. II), we deduce that R^{ss}(resp. R^s) is an open subset of R.

Then by the local universal property of R (cf. (iii) of Prop. 6, 2, Chap. II) it follows immediately that T^{ss}(resp. T^s) is open in T.

Corollary 2. Let $\chi : R^{ss} \longrightarrow H^N_{p,r}(E)^{ss}$ be the canonical morphism induced by the multivalued set mapping $\Phi : Q \longrightarrow H^N_{p,r}(E)$ as in the above proposition.

Then if m and N is sufficiently large, χ is a proper morphism; in fact we can find an integer m_0 and an ordered set of points $\mathfrak{n}$ on X such that for any integer m and an ordered set of points $\mathfrak{n}$ with $m \geq m_0$ and $\mathfrak{n} \supset \mathfrak{n}_0$, if $\chi : R^{ss} \longrightarrow H^N_{p,r}(E)^{ss}$ is the canonical associated morphism, then χ is a closed immersion.

Proof. We shall prove first that χ is proper when m and N are sufficiently large.

Let us denote by the same letter Φ the graph of the multivalued set mapping $\Phi : Q \longrightarrow H^N_{p,r}(E)$.

Let Γ be the graph of the morphism $\chi : R^{ss} \longrightarrow H^N_{p,r}(E)^{ss}$ and Ψ the closure of Γ in $Q \times H^N_{p,r}(E)$.

We have $\Phi \supset \Psi$ (see the discussion preceding the above proposition).

By the above proposition, we have (for sufficiently large m and N)

$$\Phi \cap (Q \times H^N_{p,r}(E)^{ss}) = \Gamma .$$

It follows then that

$$\Psi \cap (Q \times H^N_{p,r}(E)^{ss}) = \Gamma .$$

Since Ψ is closed in $Q \times H^N_{p,r}(E)$, the above relation implies that Γ, which by the definition is closed in $R^{ss} \times H^N_{p,r}(E)^{ss}$ is in fact closed in $Q \times H^N_{p,r}(E)^{ss}$.

Since Q is projective in particular proper over $\mathcal{C}$, the canonical projection of $Q \times H^N_{p,r}(E)^{ss}$ onto $H^N_{p,r}(E)^{ss}$ is proper and this implies that

$$\chi : R^{ss} \longrightarrow H^N_{p,r}(E)^{ss}$$

is proper for sufficiently large m and N.

Finally to show that χ is in fact a closed immersion for a choice of m and n as indicated above, it suffices to show that there exists n_0 such that whenever $n \supset n_0$, χ is an immersion.

This is quite easy and in fact a consequence of a more general fact.

Let us show for example that there exists n such that whenever $n \supset n_0$, χ is injective.

Now $\{F_q\}_{q \in R}$ represents a family of quotient bundles of E such that if $q_1 \neq q_2$, then the canonical maps $E \longrightarrow F_{q_1}$, $E \longrightarrow F_{q_2}$ represent distinct quotient bundles of E.

It follows then easily that given distinct points $q_1, \ldots, q_s$ of R, there exists a point $P \in X$ such that the fibres of $F_{q_1}, \ldots, F_{q_s}$ at the point.

P considered canonically as quotient spaces of $H^0(E)$ are distinct.

From this one deduces easily (by the well-known diagonal argument) that there exists an ordered set of points n_0 such that the canonical morphism $\chi_1 : R \rightarrow H^N_{p,r}(E)$ induced by Φ (associated to n_0) is injective.

This proves the assertion regarding injectivity.

To show there exists n_0 such that $\chi : R^{ss} \longrightarrow H^N_{p,r}(E)$ is an immersion, we have only to show that the differential map $d\chi$ of χ is injective at the tangent space of every one of the points of R^{ss} (note that R and $H^N_{p,r}(E)$ are smooth).

The proof of this is similar and is left to the reader.

This completes the proof of the corollary.

5 . Proof of the main results.

We say that a vector bundle V on X is unitary (resp. irreducible unitary) if it is associated to a unitary (resp. irreducible unitary) representation of Γ_o.

We say that a π -vector bundle on X is π -unitary (resp. irreducible π -unitary) if it is associated to a unitary representation of Γ .

(We recall that $p_o : \tilde{X} \longrightarrow X$ represents a simply connected covering of X, $\Gamma_o = \pi_1(X)$, $Y = X/\Gamma$ and $\pi = \Gamma/\Gamma_o$).

Proposition 10. (i) Let V be a unitary (resp. π -unitary vector bundle on X).

Then V is semi-stable (resp. π -semi-stable).

(ii) Let V be an irreducible unitary (resp. irreducible π -unitary) vector bundle on X.

Then V is stable (resp. π -stable).

(iii) Let $\{V_t\}_{t \in T}$ be an analytic family of vector bundles (resp. π -vector bundles) on X such that V_t is stable (resp. π -stable) of degree zero for every $t \in T$.

Then the subset T_o of T formed by points $t \in T$ such that V_t is unitary (resp. π -unitary) is a closed subset of T (of the underlying topological space).

Proof. (i) To prove (i) it suffices to show that if V is unitary (the π -unitary case is a consequence of this), then V is semi-stable.

Suppose that V is not semi-stable.

Then there exists a sub-bundle W of V such that $\deg W > 0$.

Let $k = \mathrm{rk}\ W$. Then $\overset{k}{\wedge} W$ is a line bundle such that $\deg(\overset{k}{\wedge} W) = \deg W > 0$ and $\overset{k}{\wedge} W$ is a line sun-bundle of $\overset{k}{\wedge} V$.

Now $\overset{k}{\wedge} V$ is again unitary so that we can suppose without loss of generality that W is a line bundle.

We can find a line bundle W_1 such that $\deg W_1 = \deg W$ and W_1 has a section s, $s \neq 0$, vanishing at least at one point of X (for W_1 defined by a divisor D with support at a unique point of P and multiplicity at $P = (\deg W)$.

Then we have $W_1 = W \otimes L$ where L is a line bundle of degree zero.

Therefore L is unitary (we use here the classical theorem that a line bundle on X of degree zero is unitary in fact this is also a consequence of the next theorem) so that $V \otimes L$ is also unitary.

Thus again to prove (i), we can suppose without loss of generality that W has a section s, $s \neq 0$, vanishing at least at one point of X.

But V has only <u>constant</u> sections (i.e. if E is the representation space of a representation of Γ_o defining V, V is associated to the canonical Γ_o-bundle $\tilde{X} \times E$ and Γ_o-sections of this bundle on $\tilde{X}$ are constant i.e. given by Γ_o-invariant elements of E) by Prop. 1, § 1, Chap. I

This leads to a contradiction.

This proves (i).

(ii) Let V be an irreducible unitary (resp. irreducible π-unitary vector bundle on X.

Then by (i) V is semi-stable (resp. π-semi-stable).

Suppose that V is not stable (resp. π-stable).

Then there is a sub-bundle (resp. π-sub-bundle) W of V such that deg W = 0 .

Let k = rk W.

Then similar to what we did in (i), we can suppose without loss of generality that $\overset{k}{\wedge} W$ is trivial (resp. trivial as a π-bundle, for this we make use of the fact that a π-line bundle L such that π-deg L = 0, is unitary.

This will also follow from the next theorem).

Then there is non-zero section (resp. π-section) s_o of $\overset{k}{\wedge} W$.

Now s can be identified canonically with a section (resp. π-section) of $\overset{k}{\wedge} V$ and is denoted by s.

Since $\overset{k}{\wedge} W$ is a line bundle, for every $x \in X$, $s_o(x)$ is a decomposable element of the fible of $\overset{k}{\wedge} W$ at x.

Therefore s(x) is again a decomposable element of the fibre of $\overset{k}{\wedge} V$ at x.

But $\overset{k}{\wedge} V$ being unitary (resp. π-unitary), s is a constant section (in the sense explained in (i) above).

This implies easily that there exist sections (resp. π-sections) $s_1, \ldots ; s_k$ of V such that

$$s = s_1 \wedge \ldots \wedge s_k$$

But $s_i = 0$ since V is irreducible unitary (resp. irreducible π-unitary) so that s = 0.

This leads to a contradiction.

Therefore V is stable (resp. π-stable).

This proves (ii).

(iii) Let $R = \text{rk } V_t$ for every $t \in T$.

Let S be the C-analytic space of all representations of Γ_o into GL(r, C) (resp. of all representations of Γ into GL(r, C)).

Let U denote the subset of S corresponding to unitary representations.

We see easily that there is an analytic family $\{W_s\}_{s \in S}$ of vector bundles (resp. π—vector bundles) on X parametrized by S.

Let K be the subset of $S \times T$ consisting of point (s, t) such that Hom $(W_s, V_t) \neq 0$ (resp. Hom $(W_s, V_t) \neq 0$).

By the semi-continuity theprem, K is closed in S T.

Since U is compact, $K_o = pr_T(K \cap (U \times T))$ is closed in T.

Now $t \in K_o$ if and only there is a unitary (resp. π-unitary) bundle W_s, $s \in U$ such that Hom $(W_s, V_t) \neq 0$.

Since W_s is semi-stable (resp. π-semi-stable) and V_t is stable (resp. π-stable), this is equivalent to saying $W_s \approx V_t$ (cf. Prop. 2, § 1) i.e. $K_o = T_o$.

Therefore T_o and (iii) is proved.

This proves the proposition.

Theorem 4. (i) Let V be an irreducible π-unitary bundle on X (X being a smooth complete curve over $\mathscr{C}$ on which a finite group operates); let τ be its local type.

Then every π-stable bundle W on X which is locally of type τ (cf. Remark 1, Prop. 2, Chap. I) or equivalently locally isomorphic to V (i.e. for every $x \in X$, π_x-isomorphic to V is some π_x-isomorphic to V in some $_x$-invariant neighbourhood of x) is also π-unitary.

(ii) Every π-line bundle V on X such that π-deg V = 0 (this is equivalent to assuming deg V = 0, cf. definition of π-degree given before Pro. 3, Chap. I). is -unitary.

(iii) Suppose that h = genus of Y $\geq$ 2, Y = X mod π

Then every π-stable vector bundle on X is $\overset{\circ}{\pi}$-unitary

Proof. Choose an integer m such that $H^o(W(m))$ generates W(m) and $H^1(W(m)) = 0$ for every $W \in \mathcal{S}_r$ - the category of semi-stable vector bundles on X of rank r, where r = rk V.

By Prop. 10, V is π-stable so that we have also that $H^o(V(m))$ generates V(m) and $H^1(V(m)) = 0$.

Let R = R(E/P0, P being the Hilbert polynomial of V(m), (notations being as in Prop. 6, Chap. II) and E the π-vector bundle on X associated to the canonical π-module $H^o(V(m))$.

Then the equivalent π-vector bundle E $\longrightarrow$ V(m) is in R^{τ}

Now R^{τ} is irreducible and smooth (cf. Prop. 6, Chap. II).

let $R^{\tau,s}$ denote the subset of R^{τ} of points q R such that the π-vector bundle F_q is π-stable.

Then $R^{\tau,s}$ is a Zariski open subset of R by Cor. 1, Prop. 9, Chap. II. Therefore $R^{\tau,s}$ is also irreducible, in particular connected.

By Prop. 7 and Prop. 10 (Chap. II), the subset U of points $q \in R^{\tau,s}$ such that if $F_q = V_q(m)$, V_q is π-unitary is open and closed in $R^{\tau,s}$ for the topology of the underlying analytic spaces.

Now U is non-empty by hypothesis.

Therefore $U = R^{\tau,s}$ so that we conclude that for every

$q \in R^{\tau, s}$ if $F_q = V_q(m)$, then V_q is π-unitary.

Since every π-stable vector bundle W of π-degree zero and of local type τ occurs in $R^{\tau, s}$, the part (i) of the theorem follows.

(ii) and (iii). The group Γ ($\tilde{X}/\Gamma = Y$, $\tilde{X}$ a simply-connected covering of X) is generated by elements $A_1, B_1, \ldots A_h, B_h, C_1, \ldots, C_m$ with the relations

$$A_1 B_1 A_1^{-1} B_1^{-1} \ldots . A_h B_h A_h^{-1} B_h^{-1} C_1 \ldots C_m = \mathrm{Id},$$

$$C_1^{n_1} = \ldots = C_m^{n_m} = \mathrm{Id} .$$

The C_ν, $1 \leq \nu \leq m$; can be identified with generators of π_x, x_ν being a point of X chosen over every ramification point y_ν of $p : X \to Y$.

Giving local representations $\rho_\nu : \pi_{x_\nu} \longrightarrow GL(r)$ is equivalent to giving matrices $\rho(C_\nu)$, $1 \leq \nu \leq m$.

Let us take the local type τ to be defined by $\rho(C_\nu)$, $1 \leq \nu \leq m$.

Suppose that there is a π-stable vector bundle V of rank r and π-degree zero and of local type τ.

Then since V is π-indecomposable (Cor. Prop. 2, Chap. II), by Th. 1, Chap. I, V is defined by a representation $\rho : \Gamma \longrightarrow GL(r)$.

Then $\rho(C_\nu)$ is conjugate to $\rho_\nu(C_\nu)$, $1 \leq \nu \leq m$.

We have

$$\prod_{\nu=1}^{m} \det \rho (C_\nu) = Id$$

since

$$\prod_{1 \leq i \leq h} \det \rho (A_i) \det \rho (B_i) . \det \rho (A_i^{-1}) \det \rho (B_i^{-1}) = Id.$$

We get then

$$(*) \qquad \prod_{\nu=1}^{m} \det \rho_\nu (C_\nu) = Id .$$

If $h \geq 2$, by Prop. 8, Chap. I we get then that there is an irreducible unitary representation ρ of Γ such that $\rho (C_\nu) = \rho_\nu (C_\nu)$.

This shows that if h = genus of Y is ≥ 2 given a π-stable vector bundle V of local type τ, there is a π-irreducible unitary vector bundle of local type τ.

Then by (i), we conclude that every π-stable vector bundle is π-unitary.

This proves (iii).

The condition ($*$) in the case $r = 1$ is simply

$$\prod_{\nu=1}^{m} \rho_\nu (C_\nu) = 1$$

Now given complex numbers such that

$$\alpha_\nu^{n_\nu} = 1, \; 1 \leq \nu \leq m \quad \text{and} \quad \prod_{\nu=1}^{m} \alpha_\nu = 1$$

there is obviously a unitary character $\rho : \Gamma \longrightarrow GL(1)$ such that $\rho(C_\nu) = \alpha_\nu$.

This again proves because of (i), that every π-line bundle V such that π-deg $V = 0$ is π-unitary.

This completes the proof of the theorem.

<u>Theorem 5.</u> Suppose that h = genus of $Y \geq 2$, $Y = X \bmod \pi$, X being a smooth complete curve over C on which a finite group operates.

Let $\underset{\sim}{S}_\tau$ denote the category of π-<u>semi-stable</u> vector bundles on X of rank r, degree zero and <u>fixed local type</u> τ (for definition of local type cf. Remark 1, Prop. 2, Chap. I).

Let S_τ denote the set of equivalence classes of objects in $\underset{\sim}{S}_\tau$ under the equivalence relation $V_1 \sim V_2$, V_1, $V_2 \in \underset{\sim}{S}_\tau$ if and only if $gr_\pi V_1 = gr_\pi V_2$, (see the definition inclused before Prop. 2, § of chapter II).

Then on S_τ there is a unique structure of a <u>normal projective variety</u> denoted again by S_τ such that (i) if $\{V_t\}_{t \in T}$ is an algebraic family of π-vector bundles such that $V_t \in S_\tau$ for every $t \in T$ then the canonical set map $f : T \rightarrow S$ defined by $t \overset{}{\underset{S}{\longmapsto}} gr_\pi V_t$ is a morphism and (ii) given another structure S' on having the property (i) the canonical set map $S_\tau \longrightarrow S'$ is a morphism.

Further the underlying topological space of S_τ coincides canonically with the topological space of equivalence classes of π-unitary vector bundles on X of local type τ.

Proof. With respect to the category $\mathcal{S}_{mr}$ xhoose an integer m so that $H^o(V(m))$ generates $V(m)$ and $H^1(V(m)) = 0$ for every $V \in \mathcal{S}_r$.

Let E be the π-vector bundle on X associated to the canonical π-module $H^o(V(m))$, $V \in \mathcal{S}_{m\tau}$.

Let $H = \mathrm{Aut}_\pi E$ and $G = \mathrm{Aut}\, E$ i.e. the group of automorphism of the underlying trivial vector bundle of E.

Let P be the Hilbert polynomial of $V(m)$, $V \in \mathcal{S}_{mr}$.

Let $R = R(E/P)$ and R^τ ($\subset R$) be the smooth schemes on which G and H respectively operate as in Remark 2, Prop. 6, Chap. II.

Let R^{ss} (resp. $R^{\tau, ss}$) denote the set of $q \in R$ (resp. R^τ) such that if $F_q = V_q(m)$ then $V_q \in \mathcal{S}_{mr}$ (resp. $\mathcal{S}_{m\tau}$).

Let $p = \mathrm{rk}\, E$.

Then for a suitable choice of m and an ordered set n of points $P_1, \ldots, P_N$, the canonical G-morphism

$$\chi : R^{ss} \longrightarrow H^N_{p,r}(E)^{ss}$$

is a closed immersion (by Cor. 2, Prop. 9, Chap. II).

Now $H^N_{p,r}(E)^{ss}$ has a good quotient modulo G and this is projective (by Theorems 2 and 3 of Chap. II).

Now χ being a closed immersion R^{ss} has also a good quotient $\theta : R^{ss} \longrightarrow M$ modulo G and M is projective (by Prop. 8 of Chap. II).

Now $H \subset G$ and H being a reductive subgroup a point

$x \in H^N_{p,r}(E)^{ss}$ is also semi-stable for the canonical linear action of SH-the subgroup of H of determinant one (considered as elements of GL(r)).

Now the canonical morphism $R^{\tau, ss} \longrightarrow H^N_{p,r}(E)^{ss}$ induced by χ is an H-morphism and is a closed immersion.

Now $H^N_{p,r}(E)^{ss}$ has a good quotient modulo H which is projective and so again $R^{\tau, ss}$ has a good quotient $\theta : R^{\tau, ss} \longrightarrow N$ modulo H and N is projective.

Since good quotients are categorical quotients (by Prop. 8, § 3) and R^{ss}, $R^{\tau, ss}$ are smooth varieties in particular normal, it follows easily that M and N are normal varieties.

We shall now show that N can be identified canonically as a set with S_τ and that the structure of a variety on S_τ induced by N satisfies the requirements of the theorem.

Let U denote the set of unitary representations of local type τ.

As we have often done before, using an analytic family of π-vector bundles containing all π-vector bundles corresponding to U and the local universal property of R^τ (cf. Prop. 6, Chap. II), we ger a canonical continuous map $\lambda : U \longrightarrow N$.

By the previous theorem (Th. 4, Chap. II), it follows that λ is <u>surjective</u>.

Further, if u_1, u_2 U and the representations of corresponding to u_1, u_2 respectively are equivalent, then $\lambda(u_1) = \lambda(u_2)$.

Suppose we show that

$$q_1, q_2 \in R^{\tau, ss}, \ gr_{\pi} F_{q_1} \neq gr_{\pi} F_{q_2}$$

(✱)

then $\theta_{\tau}(q_1) \neq \theta_{\tau}(q_2)$.

Then it will follow that λ induces a homeomorphism of the topological space of equivalence classes of unitary representations of Γ of rank r and local type τ and in fact that $\theta_{\tau}(q_1) = \theta_{\tau}(q_2)$ if and only if $gr_{\pi} F_{q_1} = gr_{\pi} F_{q_2}$.

The universal properties of the theorem are now immediate consequences of the local universal property of R^{τ} given by Prop. 6, Chap. II.

Thus to include the proof of the theorem it suffices to prove (✱).

To prove (✱), we have to ahow that if $q_1, q_2 \in R^{\tau, ss}$ as in (✱), then

$$\overline{0(q_1)} \cap \overline{0(q_2)} \quad \text{is empty}$$

where $\overline{0(q_1)}$, $\overline{0(q_2)}$ represent the closures in $R^{\tau, ss}$ of the orbits $0(q_1)$, $0(q_2)$ through q_1, q_2 respectively under H.

Suppose that $q \in \overline{0(q_1)} \cap \overline{0(q_2)}$.

Then we can find a smooth curve C and a morphism $\mu : C \longrightarrow \overline{0(q_1)}$ such that if ξ is the generic point of C then $\mu(\xi) \in 0(q_1)$ and there is a (closed) point $\xi_o \in C$ such that

$\mu(\xi_o) = q$.

We can now suppose without loss of generality that $gr_{\pi} F_y$ is constant for every y of the form $\mu(x)$, $x \in C$, $x \neq \xi_o$ (by taking C to be a suitable neighbourhood of ξ_o).

Taking the inverse of the family $\{F_q\}$ by the morphism μ to prove (✱), we are easily reduced to prove the following

Lemma 7. Let $\{V_t\}_{t \in T}$ be an algebraic family of vector bundles (resp. π-vector bundles) on X parametrized by an irreducible smooth curve T and to a point of T.

Suppose that (i) $\forall$ t, V_t is semi-stable of degree zero (resp. π-semi-stable of π-degree zero) and (ii) $gr\ V_t$ (resp. $gr\ V_t$) is constant for $t \in T$, $t \neq t_o$.

Then $gr\ V_t = gr\ V_{t_o}$ (resp. $gr_{\pi}\ V_t = gr_{\pi}\ V_t$) for any t T.

Proof of lemma. Let $gr\ V_t = W_1 \oplus \dots \oplus W_k$, W_i stable of degree zero (resp. π-stable of π-degree zero) for $t \in T$, $t \neq t_o$.

The proof is by induction on k.

Let $k = 1$.

Let $D = \{ t \in T \mid \dim \mathrm{Hom}\,(W_1, V_t)$ (resp. $\dim \mathrm{Hom}_{\pi}(W_1, V_t)) > 0 \}$.

By the semi-continuity thorem D is closed in T.

But $S \supset T_o$ where $T_o = T - t_o$.

Therefore $D = T$.

This implies that $W_1 \approx V_t$ (cf. Prop. 2, § 1, Chap. II).

This proves the lemma for the case $k = 1$.

Let us go th the general case.

Let $D_1 = \{ t \in T \mid \dim \mathrm{Hom}\,(W_1, V_t\ )$

(resp. dim Hom $_{\pi}$ $(W_1, V_t)) > 0 \}$.

Then D_1 is closed in T and $\bigcup_{\ell} D_{\ell} = T$.

This implies that at least one D_{ℓ} is T; say $D_1 = T$, so that we are have Hom (W_1, V_t) (resp. Hom$_{\pi}$ $(W_1, V_t) \neq 0$. for every t T.

We can now suppose that dim Hom (W_1, V_t) (resp. Hom (W_1, V_t)) is constant for every $t \in U$ where U is a non-empty open subset of T.

Let V be the defining bundle (resp. π -bundle) on $X \times T$ of the family V_t .

Then $P = (pr_T)_* \; (Hom\, ((pr_X)^* \; (W_1), V)) \neq 0$ where Hom $((pr_X)^* (W_1), V))$ denotes the sheaf of germs of homomorphisms of $(pr_X)^* (W_1)$ into V.

We can suppose without loss of generality that T is an affine nighbourhood of T .

Then the above implies that there is a homomorphism $f : (pr_X)^* \; (W_1) \longrightarrow V$ and we can suppose without loss of generality that the restriction of f to every $X \times t$, $t \times T$, is non-zero (for $P = P_1 \oplus P_2$, where P_1 is locally free and P_2 is a torsion sheaf and we can take a section of P which comes from that of P_1 and does not vanish on the fibre of the vector bundles associated to P_1 at t_o).

This implies that $(pr_X)^* \; (W_1)$ can be considered as a sub-bundle of V; let V_1 be the quotient bundle of V by this sub-bundle.

Now the family defined by V_1 satisfies the hypothesis of the lemma and $(gr\ V_1)_t$ (resp. $(gr_{\pi}\ V_1)_t$) is of length $< k$.

Now the induction works and the lemma is proved.

This completes the proof of the theorem.

Remarks (i) Taking for π the trivial group reducing to one element, we get the theorem for semi-stable vector bundles of fixed rank and degree zero on a smooth complete curve of genus ≥ 2, proved in [17] .

Note that the above theorem (Th. 5) because of Remark, Prop. 5, Chap. I includes also the generalisations of the main theorem of 17 to arbitrary degree.

(ii) The dimension of the variety S_{τ} (when S_{τ} is not empty is

$$r^2(h-1) + 1 + \frac{1}{2} \sum_{\nu=1}^{m} e_{\nu}$$

where h = genus of Y, r = rank of the vector bundles in question and e_{ν} the integers determined by the local type τ (cf. (v), Prop. 6, Chap. II and Prop. 7, Chap. I).

(iii) The variety S_{τ} is smooth at the points corresponding to irreducible π-unitary vector bundles or equivalently π-stable vector bundles.

This is an immediate consequence of Cor., Prop. 6, Chap. II.

REFERENCES

[1] M.F. Atiyah, Vector bundles over an elliptic curve. Proc. London Math. Soc., Third Series, 7(1957), 412-452.

[2] H. Cartan, Quotient d'un espace analytique par un groupe d'automorphismes, Algebraic geometry and topology, A symposium in honour of S. Lefschetz , Princeton University Press, 1957 (Princeton Math. Series, n).

[3] H. Cartan and S. Eilenberg, Homological algebra, Princeton University Press, 1956.

[4] A. Grothendisck, Sur la mémoire de Weil "Généralisation des fonctions abeliennes, Seminaire Bourbaki, expose 141, (1956-57).

[5] ———, Sur quelques points d'algèbre homologique, Tohoku Math. J., Series 2, 9 (1957), 119-121.

[6] ———, Les schémas de Hilbert, Séminaire Bourbaki, expose 221, t. 13, 1960-61.

[7] ———, et J. Dieudonné, Elements de Géometrie algébrique, Publ. Math., Inst. Hautes Etudes Scientifiques.

[8] D. Mumford, Projective invariants of projective structures and applications. Proc. Intern. Cong. Math., Stockholm, 1962, 526 - 530.

[9] ———, Geometric invariant theory, Springer-Verlag, 1965.

[10] ———; Abelian varieties (to appear), Oxford University Press, Studies in Mathematics, Tata Institute of Fundamental Research.

[11] M.S. Narasimhan and C.S. Seshadri, Holomorphic vector bundles on a compact Riemann surface, Math. Ann., 155 (1964), 69-80.

[12] ———, Stable and unitary vector bundles on a compact

Riemann surface, Ann. of Math. 82 (1965), 540-567.

[13] A. Selberg, On discontinuous groups in higher dimensional symmetric spaces, Contributions to function theory, Tata Institute of Fundamental Research, 1960.

[14] J. - P. Serre, Faisceaux algebriques cohérents, Ann. of Math., 61 (1955), 197-278.

[15] ________________, Géometrie analytique et géometric algébrique, Ann. Inst. Fourier, 1955—56, 1 - 42.

[16] C. S. Seshadri, Generalized multiplicative meromorphic functions on a complex analytic manifold, J. Indian Math. Soc. 21 (1957), 149 - 178.

[17] ________________, Space of unitary vector bundles on a compact Riemann surface, Ann of Math. 82 (1965), 303-336.

[18] ________________, Mumford's conjecture for GL(20) and applications, Algebraic geometry, Bombay Colloquium (1968), Oxford University Press.

[19] A. Weill, Généralisation des fonctions abeliennes, J. Math. pures appl. 17 (1938), 47 - 87.

[20] ________________, Remark on the cohomology of groups, Ann. of Math., 80 (1964), 149. - 157.

CENTRO INTERNAZIONALE MATEMATICO ESTIVO

(C. I. M. E.)

O. ZARISKI

CONTRIBUTIONS TO THE PROBLEM OF EQUISINGULARITY

Corso tenuto a Varenna dal 7 al 17 Settembre 1969

O. Zariski

CONTENTS.

CONTRIBUTIONS TO THE PROBLEM OF EQUISINGULARITY

by

Oscar Zariski

Harvard University, U.S.A.

INTRODUCTION. The general problem which we propose in these lectures is the following: given an irreducible subvariety W of the singular locus of an algebraic (or algebroid) variety V and given a simple point Q of W, give a precise meaning to the following intuitive statement: "the singularity which V has at the point Q is 'not worse' than (or is 'of the same type' as) the singularity which V has at the general point of W". We briefly phrase this statement as follows: "V is equisingular along W, at Q." It is understood that we require the solution to consist not merely of some plausible definition and some reasonable consequences, but above all of a body of criteria of various nature (algebro-geometric, topological and differentio-geometric) and of the proofs of equivalence of these various criteria.

In this lecture we give, in the first place, a complete solution of this problem in the special case of $\mathrm{cod}_V W = 1$. We also treat a special type of equisingularity which we call equisaturation; we are led to this concept by our algebraic theory of saturation and saturated local rings. For both of these topics we need a thorough analysis of some old and new aspects of the concept of equivalent singularities of plane algebroid curves. This analysis is developed in Sections 1-6. In the last section we discuss connections with the differentiogeometric conditions A and B of Whitney-Thom.

1. Plane Algebroid Curves

We shall begin with the theory of singularities of (formal) plane algebroid curves (more precisely: algebroid curves which admit an embedding in the plane). Our object will be to present some aspects - both old and new - of the classical concept of equivalence between such singularities. We shall restrict ourselves to ground fields k which are algebraically closed and of characteristic zero.

We shall be guided by the following intrinsic definition of a plane algebroid curve C (over k) : it is a neotherian complete local ring $\mathfrak{o}$ having the following properties :

a) $\mathfrak{o}$ has Krull dimension 1.

b) $\mathfrak{o}$ has no nilpotent elements (other than zero).

c) The maximal ideal $\mathfrak{m}$ of $\mathfrak{o}$ can be generated by two elements.

d) k is a coefficient field of $\mathfrak{o}$.

We say then that $\mathfrak{o}$ is the local ring of C.

Condition (a) signifies that 1) there exist ideals $\mathfrak{q}$ primary for $\mathfrak{m}$ (i.e., which are primary and have $\mathfrak{m}$ as associated prime ideal) which are generated by a single element and that 2) the zero ideal is not primary for $\mathfrak{m}$. Any element x of $\mathfrak{o}$ such that the principal ideal $\mathfrak{o}x$ is primary for $\mathfrak{m}$ is called a parameter of $\mathfrak{o}$ (or of C). It is obvious, of course, that every parameter belongs to $\mathfrak{m}$ (hence is a non-unit). It also follows easily from 2) that a parameter of $\mathfrak{o}$ is never a zero-divisor.

Condition (b) signifies that the zero ideal is a finite (irredundant) intersection of prime ideals :

(1) $$(0) = \mathfrak{p}_1 \cap \mathfrak{p}_2 \cap \ldots \cap \mathfrak{p}_h ,$$

By (a) we must have $\mathfrak{p}_i \subset \mathfrak{m}$ (i=1, 2, ..., h), and it thus follows at once from (a) and (b) that an element x of $\mathfrak{m}$ is a parameter if and only if x is not a zero-divisor (equivalently: if and only if $x \notin \mathfrak{p}_i$, i = = 1, 2, ..., h) . The h+1 prime ideals $\mathfrak{p}_1, \mathfrak{p}_2, \ldots, \mathfrak{p}_h, \mathfrak{m}$ are all the prime ideals of $\mathfrak{O}$. Thus the scheme Spec ($\mathfrak{O}$) has exactly h+1 points; of these, the point represented by $\mathfrak{m}$ is the only closed point.

Condition (d) signifies that k is a field contained in $\mathfrak{O}$ and that the injection $k \to \mathfrak{O}/\mathfrak{m}$ induced by the canonical homomorphism $\mathfrak{O} \to \mathfrak{O}/\mathfrak{m}$ is surjective, hence is an isomorphism $k \xrightarrow{\sim} \mathfrak{O}/\mathfrak{m}$

The three conditions (a), (b) and (d) define the general concept of an algebroid curve C , over an arbitrary , algebraically closed ground field k (not necessarily of characteristic zero). We now look at condition (c) .

Let $\{x_1, x_2, \ldots, x_n\}$ be any basis of $\mathfrak{m}$. Using (d) and the completeness of $\mathfrak{O}$ we find a natural surjective k-Homomorphism

$$\Phi : k\,[[X_1, X_2, \ldots, X_n]] \to \mathfrak{O} = k\,[[x_1, x_2, \ldots x_n]] \quad ,$$

such that $\Phi(X_i) = x_i (i=1, =, \ldots, n)$; here $k\,[[X_1, X_2, \ldots, X_n]]$ denotes the ring of formal power series in the indeterminates X_i , with coefficients in k. Thus

$$\text{(2)} \qquad \mathfrak{O} = k\,[[x_1, x_2, \ldots, x_n]] = k\,[[X_1, x_2, \ldots, X_n]]/\mathfrak{A} \,,$$

where $\mathfrak{A}$ is an ideal in $k\,[[X_1, X_2, \ldots, X_n]]$. Conditions (a) and (b) signify that $\mathfrak{A}$ is a finite intersection of h one-dimensional prime ideals in $k\,[[X_1, X_2, \ldots, X_n]]$, where h is the integer which occurs in (1). We thus have an embedding of C as an algebroid curve in the affine n-space (over k). Of particular interest are those embeddings of C which are associated with minimal bases of $\mathfrak{m}$ (all minimal bases of $\mathfrak{m}$ have the same number of elements; this number is the dimension of the k-vector space $\mathfrak{m}/\mathfrak{m}^2$, and is sometimes denoted by Emb $\mathfrak{O}$ and is called the embedding dimension of $\mathfrak{O}$, or of C).

Condition (c) signifies that the embedding dimension of C is 1 or 2. If n=1, i.e., if $\mathfrak{m}$ is a principal ideal, the local ring $\mathfrak{O}$ is regular, C is a regular branch; equivalently: $\mathfrak{O}$ is a formal power series ring in one indeterminate, and - correspondingly - C, for a suitable embedding, is the affine line. We exclude this case, and we assume therefore that the embedding dimension of C is 2. If we then take a minimal basis $\{x_1, x_2\}$ of $\mathfrak{m}$, the power series ideal $\mathfrak{U}$ in (2) will be a principal ideal: $\mathfrak{U} = (f(X_1, X_2))$, and f a product of h distinct (non-associated) irreducible power series:

(3) $$f = f_1 f_2 \ldots f_h .$$

We then write symbolically

(4) $$C : f(X_1, X_2) = 0 ,$$

and we say that (4) is an equation of C. The power series f is uniquely determined by the basis $\{x_1, x_2\}$ of $\mathfrak{m}$, up to an arbitrary unit factor in $k[[X_1, X_2]]$. The transition from one minimal basis $\{x_1, x_2\}$ to another minimal basis $\{x_1', x_2'\}$ is effected by an analytic transformation of the form

$$x_i' = a_{i2}x_1 + a_{i2}x_2 + \text{terms of degree} > 1 \text{ in } x_1, x_2, \quad i = 1, 2,$$

where $a_{ij} \in k$ and $a_{11}a_{22} - a_{12}a_{21} \neq 0$.

We denote by F the total ring of quotients of $\mathfrak{o}$, and we write $F = k((C))$. If M denotes the complement of $\bigcup_{i=1}^{h} \mathfrak{p}_i$ in $\mathfrak{o}$ (i.e., if M is the set of elements of $\mathfrak{o}$ which are not zero-divisors), then F is the quotient ring $\mathfrak{o}_M$ of $\mathfrak{o}$ with respect to the multiplicative system M. We shall use freely properties of quotient rings with respect to multiplicative systems, and we refer for that to Zariski-Samuel, Commutative Algebra, Vol. I, pp. 221-231. Thus, for example, using an appropriate property of quotient rings (namely, Theorem 16, p. 224, loc. cit.), we see that $\mathfrak{o}$ is a subring of F, that the ideals $\mathfrak{f}_i = F.\mathfrak{p}_i$ are the only prime ideals of F, that these ideals are therefore maximal, and that $(0) = \mathfrak{f}_1 \cap \mathfrak{f}_2 \cap \ldots \cap \mathfrak{f}_h$, with $\mathfrak{f}_i \cap \mathfrak{o} = \mathfrak{p}_i$. If we set $F_i = \mathfrak{f}_1 \cap \mathfrak{f}_2 \cap \ldots \cap \mathfrak{f}_{i-1} \cap \mathfrak{f}_{i+1} \cap \ldots \cap \mathfrak{f}_h$, then each F_i is a field:

$$F_i \cong F/\mathfrak{f}_i,$$

and F is the direct sum of the F_i:

$$F = F_1 \oplus F_2 \oplus \ldots \oplus F_h.$$

For a given equation (4) of C we have, using the factorization (3) of f, that F is the set of all quotients $A(x_1, x_2)/B(x_1, x_2)$ such that $B(X_1, X_2)$ is not divisible by $f(X_1, X_2)$ (here $A(X_1, X_2)$, $B(X_1, X_2)$ are formal power series); $\mathfrak{p}_i = \mathfrak{o} f_i(x_1, x_2)$, $\mathfrak{f}_i =$

$= F. f_i(x_1, x_2)$, F_i = set of all above quotients $A(x_1, x_2)/B(x_1, x_2)$ such that $A(X_1, X_2)$ is divisible by $f(X_1, X_2)/f_i(X_1, X_2)$. Finally $F_i \cong$ field of fractions of the integral domain $k[[X_1, X_2]]/(f_i(X_1, X_2))$.

If h = 1, i.e., if F is a field, then we say that C is <u>irreducible</u>, or that C is an <u>algebroid branch</u>. If h > 1, we denote by γ_i the algebroid branch whose local ring is $\mathfrak{o}/\mathfrak{p}_i (i=1, 2, \ldots, h)$, and we refer to $\gamma_1, \gamma_2, \ldots, \gamma_h$ as <u>the branches of</u> C. If $\mathfrak{m} = (x_1, x_2)$, then the maximal ideal of $\mathfrak{o}/\mathfrak{p}_i$ is generated by the $\mathfrak{p}_i$ -residues of x_1, x_2, and with respect to the plane embedding of γ_i thus obtained, $f_i(X_1, X_2) = 0$ is an equation of γ_i.

It is to be observed that while the embedding dimension of any γ_i is $\leqslant 2$, it may very well happen that any given γ_i may be regular.

The multiplicity $m(C)$ of C (at its origin 0, i.e., at the point $X_1 = X_2 = 0$) can be defined intrinsically as the multiplicity $e(\mathfrak{o})$ of the local ring $\mathfrak{o}$ (see [10], volume 2, p. 294). If $s = m(C)$, then it can be seen that <u>s</u> is the degree of the initial form of f: $f = f_s + f_{s+1} + \ldots$, where f_i is a form of degree i, and $f_s \neq 0$. From (3) it follows that $m(C) = \sum_{i=1}^{h} m(\gamma_i)$. We have $m(C) = 1$ if and only if C is a regular branch.

We have the well-known concept of the graded ring $G(\mathfrak{o})$ of $\mathfrak{o}$ (see [10], volume 2, p. 249):

$$G(\mathfrak{o}) = \sum_{n=0}^{\infty} \mathfrak{m}^n / \mathfrak{m}^{n+1}$$

It follows easily that

$$G(\mathfrak{o}) \cong k[X_1, X_2] / (f_s(X_1, X_2)),$$

i.e., geometrically speaking $G(\mathfrak{O})$ is a set of <u>s</u> (not necessarily distinct) lines in the plane passing through the origin 0 of C. These are called the <u>tangent lines</u> of C.

<u>It C is irreducible, then it has only one tangent line; equivalently: if the power series f is irreducible then the initial form f_s is the power $(a_1X_1 + a_2X_2)^s$ of a linear form</u> $(a_1, a_2, \varepsilon k)$. We sketch the proof (which is based on the Weierstrass preparation theorem and on Hensel's lemma). We may assume $s > 1$. Since f is irreducible, f is not divisible by X_1, hence f is regular in X_2. By the Weierstrass preparation theorem we may therefore assume that f is a monic polynomial in X_2 with coefficients in $k[[X_1]]$. The quotient $f(X_1, X_2'X_1)/X_1^s$ is a monic polynomial $g(X_1, X_2')$ in X_2', with coefficients in $k[[X_1]]$, and is irreducible in $k[[X_1]][X_2']$. We have $g(0, X_2') = f_s(1, X_2')$, and hence, by Hensel's lemma, we must have $f_s(1, X_2') = (X_2' - \alpha)^s$, $\alpha \in k$.

q.e.d.

Since the leading form f_s of f is the product of the leading forms of the irreducible factors $f_1, f_2, \ldots, f_h$ of f, the set of tangents of C coincides with the set of tangents of the branches of C.

The ring $\mathfrak{m}/\mathfrak{m}^2$ is a vector space over k, of dimension 2. For each element x of $\mathfrak{m}$ we denote by $\bar{x}$ the associated vector in $\mathfrak{m}/\mathfrak{m}^2$ ($\bar{x} = \mathfrak{m}^2$ - residue of x). If $\mathfrak{m} = \mathfrak{O}x_1 + \mathfrak{O}x_2$, then $\bar{x}_1, \bar{x}_2$ form a basis of $\mathfrak{m}/\mathfrak{m}^2$, and thus $\bar{x}$ is a linear form $a_1\bar{x}_1 + a_2\bar{x}_2$, where $a_1 = a_2 = 0$ if and only if $x \in \mathfrak{m}^2$. We say that x is a <u>transversal parameter of</u> $\mathfrak{O}$ if $x \in \mathfrak{m}$, $x \notin \mathfrak{m}^2$ <u>and</u> if the image of x in the graded algebra $G(\mathfrak{O})$ (i.e., if the element $\bar{x}$, regarded as a homogeneous element of degree 1 of $G(\mathfrak{O})$) is not a zero-divisor in $G(\mathfrak{O})$. It is seen at once that x is a transversal parameter if and only if the linear form $a_1X_1 + a_2X_2$ (where $a_1\bar{x}_1 + a_2\bar{x}_2 = x$) is not a factor of the initial form f_s. In other words: x is a transversal parameter if and only if the vector $\bar{x}$ is not a tangent of C (equivalently: the line $a_1X_1 + a_2X_2 = 0$ is not a tangent of C). It is also immediate that a transversal parameter is indeed a parameter of $\mathfrak{O}$.

If x is expressed as a power series $g(x_1, x_2)$ of the basis elements x_1, x_2 of $\mathfrak{m}$ (whence $g(0,0) = 0$), and if we denote by D the analytic cycle defined by $g(X_1, X_2) = 0$, then x is transversal if and only if the intersection multiplicity (D,C) is equal to $\mathfrak{m}(C)$. It is also easily seen that x is transversal if and only if x is a superficial element of order 1 for $\mathfrak{m}$ (see [10], volume 2, p. 285). Finally, x is transversal if and only if the multiplicity of the principal ideal $\mathfrak{O}x$ (see [10], volume 2, p. 294) is equal to the multiplicity of C.

2. The Quadratic Transform of C

We first summarize some of the properties of quotient rings which we shall use most frequently in the sequel. For details see ([10], volume 1, pp. 221-231).

Let R be a commutative ring with identity, and let M be a multiplicative system (m. s.) in R, i.e., a non-empty subset of R which is closed under multiplication and does not contain the zero. There is then a natural homomorphism $\Phi: R \to R_M$, and in this homomorphism the elements of $\Phi(M)$ are units in R_M. The quotient ring R_M is uniquely determined, up to an isomorphism, by the existence of Φ and by a suitable "universal property". The properties of the operation of quotient ring formation which we wish to summarize are the following :

(a) Φ is an injection if and only if no element of M is a zero divisor in R. In particular, no element of $\Phi(M)$ is a zero divisor in $\Phi(R)$, and thus R_M can be identified with $\Phi(R)_{\Phi(M)}$.

(b) If $\mathfrak{p}$ is a prime ideal in R, we set $\mathfrak{p}^e = R_M \Phi(\mathfrak{p})$; and if $\mathfrak{p}'$ is a prime ideal in R_M, we set $\mathfrak{p}'^c = f^{-1}\{\mathfrak{p}'\}$. The correspondence $\mathfrak{p}' \to \mathfrak{p}'^c$ sets up a (1,1) mapping of the set of all prime ideals $\mathfrak{p}'$ in R_M onto the set of those prime ideals $\mathfrak{p}$ in R which are disjoint from M, and we have

$$(\mathfrak{p}'^c)^e = \mathfrak{p}' .$$

(c) If M' is a m.s. in R such that $M' \supset M$, then $\Phi(M')$ is a m.s. in R_M, and there is a canonical isomorphism $(R_M)_{\Phi(M')} \xrightarrow{\sim} R_{M'}$; this homomorphism is compatible with the canonical homomorphisms

$$R \to R_M, \ R \to R_{M'} \ \text{and} \ R_M \to (R_M)_{\Phi(M')} .$$

(d) If $M' \supset M$, as in (c), and if every element of M' is of the form em, where e is a unit and $m \in M$, then the canonical homomorphism $R_M \to (R_M)_{\Phi(M')}$ is an isomorphism. Therefore, we shall in this case identify R_M with $R_{M'}$.

(e). A special case of (d) leads to the following conclusion: if $\mathfrak{p}'$ is a prime ideal in R_M and if $\mathfrak{p} = \mathfrak{p}'^c$, then the two quotient rings $R_{\mathfrak{p}}$ $(=R_{R-\mathfrak{p}})$ and $(R_M)_{\mathfrak{p}'}$ are canonically isomorphic. This is seen by observing that the two multiplicative systems $R_M - \mathfrak{p}'$ and $\varphi(R-\mathfrak{p})$ in R_M are in the same relation as M' and M are in (d) (with R being replaced by R_M) and $R - \mathfrak{p} \supset M$.

Here are two preliminary applications of the above properties.

A) Let M^* be a m.s. in F and let $\mathfrak{U}^*$ be the kernel of the canonical homomorphism $\Phi : F \to F_{M^*}$. If $\xi \in \sqrt{\mathfrak{U}^*}$ then $\xi^n m = 0$ for some positive integer n and for some $m \in M^*$. Hence $(\xi m)^n = 0$, and therefore $\xi m = 0$, since F has no nilpotent elements (different from zero). This shows that $\xi \in \mathfrak{U}^*, \mathfrak{U}^* = \sqrt{\mathfrak{U}^*}$, whence $\mathfrak{U}^* = \bigcap_{i \in S} \mathfrak{f}_i$, where S is a (non-empty) subset of the set of indices $\{1, 2, \ldots, h\}$. Therefore $\Phi(F) \cong F/\mathfrak{U}^* \cong \cong \bigoplus_{i \in S} F_i$, and we can identify $\Phi(F)$ with $\bigoplus_{i \in S}$ and F_{M^*} with $(\bigoplus_{i \in S} F_i)_{\overline{M}^*}$, where $\overline{M}^*$ $(= (M^*))$ is a m.s. in $\bigoplus_{i \in S} F_i$, no elements of which is a zero-divisor. But then all elements of $\overline{M}^*$ are units, showing that $F_{M^*} = \bigoplus_{i \in S} F_i \subset F$. We have shown that every quotient ring of F can be canonically identified with a subring of F, in fact with a partial direct sum of some of the fields F_i.

B) Let $\mathfrak{O}'$ be any sub-ring of F and let M' be a m.s. in $\mathfrak{O}'$. Then $\mathfrak{O}'_{M'}$ can be canonically identified with a sub-ring of $F_{M'}$, i.e., with a sub-ring of F (by A). From now on all quotient rings $\mathfrak{O}'_{M'}$ $(\mathfrak{O} \subset F)$ will be regarded (without ambiguity) as sub-rings of F. In particular, if $\mathfrak{O}' = \mathfrak{O}$ and $M' = \mathfrak{O} - \bigcup_{i \in S} \mathfrak{p}_i$, then one sees at once that $\mathfrak{O}'_M = \bigoplus_{i \in S} F_i$.

We define the quadratic transform T(C) of C intrinsically as the following (reduced) scheme :

(1) $$T(C) = \bigcup_x \operatorname{Spec}(\mathfrak{O}[x^{-1}\mathfrak{m}]) ,$$

where x ranges over the set of non-zero divisors in $\mathfrak{m}$ and where $x^{-1}\mathfrak{m}$ denotes the set of quotients z/x, $z \in \mathfrak{m}$. It is understood that the localizations of $\mathfrak{O}[x^{-1}\mathfrak{m}]$ are identified with well defined sub-rings of F, and that we identify points of $\operatorname{Spec}(\mathfrak{O}[x^{-1}\mathfrak{m}])$ and $\operatorname{Spec}(\mathfrak{O}[x'^{-1}\mathfrak{m}])$ if their local rings coincide.

Let us fix a basis $\{x_1, x_2\}$ of $\mathfrak{m}$ such that neither x_1 nor x_2 is a zero-divisor and such that x_1 is a transversal parameter of . We shall prove now that

(2) $$T(C) = \operatorname{Spec}(\mathfrak{O}[\frac{x_2}{x_1}]) .$$

For any given x in $\mathfrak{m}$ (x - not a zero divisor), we set $x'_1 = x_1/x$, $x'_2 = x_2/x$ and $\mathfrak{O}' = \mathfrak{O}[x^{-1}\mathfrak{m}]$. We have then $\mathfrak{O}' = \mathfrak{O}[x'_1, x'_2]$. We also set

$$\mathfrak{O}'_1 = \mathfrak{O}[x_1^{-1}\mathfrak{m}] = \mathfrak{O}[x_2/x_1] ,$$
$$\mathfrak{O}'_2 = \mathfrak{O}[x_2^{-1}\mathfrak{m}] = \mathfrak{O}[x_1/x_2]$$

Let $\mathfrak{p}'$ be any prime ideal in $\mathfrak{O}'$, and let $x = u_1x_1 + u_2x_2 (u_i \in \mathfrak{O})$. We have then $1 = u_1x'_1 + u_2x'_2$, and hence either $x'_1 \notin \mathfrak{p}'$ or $x'_2 \notin \mathfrak{p}$. We shall prove the following :

C) <u>If</u> $x'_i \notin \mathfrak{p}'$ <u>(where</u> i <u>is either</u> 1 <u>or</u> 2) <u>then</u>

<u>there exists a prime ideal</u> $\mathfrak{p}_i'$ <u>in</u> $\mathfrak{O}_i'$ <u>such that</u> $(\mathfrak{O}_i')_{\mathfrak{p}_i'} = \mathfrak{O}'_{\mathfrak{p}'}$.

From C) it will follow already that

(3) $$T(C) = \mathrm{Spec}(\mathfrak{O}_1') \cup \mathrm{Spec}(\mathfrak{O}_2') .$$

We shall also show that

D) <u>If</u> x_1 <u>is a transversal parameter then</u> x_1/x_2 <u>is a unit in</u> $\mathfrak{o}_2'$. In view of D), we see that in the special case in which $x = x_2$ (and therefore $x_1' = x_1/x_2$) the assumption $x_i' \notin \mathfrak{p}'$ in C) is satisfied for $i = 1$ and for any prime ideal $\mathfrak{p}'$ in $\mathfrak{O}'(=\mathfrak{O}_2')$. Hence by C), we have $\mathrm{Spec}(\mathfrak{O}_2') \subset \mathrm{Spec}(\mathfrak{O}_1')$, and thus (2) follows from (3).

To prove C), let, say, $x_1' \notin \mathfrak{p}'$. We consider in $\mathfrak{O}'$ the multiplicative system $N' = \{x_1'^i, i \geq 0\}$, and we set $\mathfrak{O}'' = \mathfrak{O}_{N'}$. Since x_1' is not a zero-divisor, we have $\mathfrak{O}' \subset \mathfrak{O}''$, and since N' and $\mathfrak{p}'$ are disjoint, it follows (by property (b)) that if we set $\mathfrak{p}'' = \mathfrak{p}'^e (= \mathfrak{O}''\mathfrak{p}')$, then $\mathfrak{p}' = \mathfrak{p}''^c (= \mathfrak{p}'' \cap \mathfrak{O}')$. Hence, if we apply property (e) to $R = \mathfrak{O}'$, $M = N'$, with $\mathfrak{p}'$ replaced by $\mathfrak{p}''$, we find that

(4) $$\mathfrak{O}'_{\mathfrak{p}'} = \mathfrak{O}''_{\mathfrak{p}''} .$$

We now observe that $1/x_1' \in \mathfrak{O}_1'$, we consider in $\mathfrak{O}_1'$ the multiplicative system $N_1' = \{1/x_1'^i, i \geq 0\}$, and we set $\mathfrak{O}_1'' = (\mathfrak{O}_1')_{N_1'}$. We see at once that $\mathfrak{O}'' = \mathfrak{O}_1''$, since $\mathfrak{O}'' = \mathfrak{O}'[1/x_1'] = \mathfrak{O}[x_1/x, x_2/x, x/x_1] = \mathfrak{O}[x_1/x, x_2/x_1, x/x_1] = \mathfrak{O}[x_1/x, x_2/x_1] = \mathfrak{O}_1'[x_1/x] = \mathfrak{O}_1''$. It follows therefore from property (e) that $\mathfrak{O}''_{\mathfrak{p}''} = (\mathfrak{O}_1')_{\mathfrak{p}_1'}$

where $\mathfrak{p}'_1$ is the prime ideal in $\mathfrak{o}'_1$ which corresponds to the ideal $\mathfrak{p}''$ in the quotient ring $(\mathfrak{o}'_1)_{N_{1'}}$ $(= \mathfrak{o}1'' = \mathfrak{o}'')$.

It remains to prove D). We may assume that f is a monic polynomial in X_1. Since x_1 is transversal, the term X_2^s is present in the initial form $f_s(X_1, X_2)$. We have $x_1 = x_2 x'_1$, and thus $f(x_2 x_1', x_2) = 0$. This equation is of the form $x_2^s g(x'_1, x_2) = 0$, where g is a polynomial in x'_1, with coefficients in $k[[x_2]]$, and where $g(0,0) \neq 0$. Since x_2 is not a zero-divisor, it follows that $g(x'_1, x_2) = 0$, showing that $\mathfrak{o}'_2 x'_1 + \mathfrak{o}'_2 x_2 = (1)$. Now, let $\mathfrak{p}'$ be any prime ideal in $\mathfrak{o}'_2$. If $\mathfrak{p}' \cap \mathfrak{o} = \mathfrak{m}$, then $x_2 \in \mathfrak{p}'$, whence $x'_1 \notin \mathfrak{p}'$. If $\mathfrak{p}' \cap \mathfrak{o} = \mathfrak{p}_i$ $(1 \leqslant i \leqslant h)$, then $x'_1 \notin \mathfrak{p}'$, since x'_1 is not a zero-divisor. This proves D).

We may assume that f is a monic polynomial in X_2. We denote by $\mathfrak{o}'$ the ring $\mathfrak{o}[x'_2]$, $x'_2 = x_2/x_1$. Since x_1 is a transversal parameter, f is exactly of degree s in X_2. Thus the relation $f(x_1, x_2) = 0$ has the form

$$x_2^s + A_1(x_1)x_1x_2^{s-1} + A_2(x_1)x_1^2x_2^{s-2} + \dots + A_s(x_1)x_1^s = 0,$$

where the $A_i(x_1)$ are power series in x_1. This equation implies that $x_2/x_1 (= x'_2)$ <u>is integral over</u> $k[[x_1]]$. Since $\mathfrak{o} = k[[x_1]][x_2]$, it follows that $\mathfrak{o}' (= \mathfrak{o}[x_2/x_1]) = k[[x_1]][x_2/x_1]$ and that consequently $\mathfrak{o}'$ <u>is a complete semi-local ring.</u> Let $\mathfrak{m}'_1, \mathfrak{m}'_2, \dots, \mathfrak{m}'_d$ be the distinct maximal ideals of $\mathfrak{o}$. In view of the integral dependence of x'_2 over $k[[x_1]]$, k is a coefficient field of each of the d <u>local</u> complete rings $\mathfrak{o}'_{\mathfrak{m}'_j}$.

We denote these $\underline{d}$ rings by $\mathfrak{O}_1', \mathfrak{O}_2', \ldots, \mathfrak{O}_d'$. If c_j is the $\mathfrak{m}_j'$ - residue of x_2' in k, then $\mathfrak{m}_j' = \mathfrak{O}' x_1 + \mathfrak{O}'(x_2'-c_j)$. Thus, the d constants $c_1, c_2, \ldots, c_d$ are distinct, and each local ring $\mathfrak{O}_j'$ has embedding dimension $\leqslant 2$. Let C_j' be the plane algebroid curve defined by the local ring $\mathfrak{O}_j'$. The origin $0_j'$ of C_j' in the (X_1, X_2') - plane is the point $X_1 = 0, X_2' = c_j$. The constants $c_1, c_2, \ldots, c_d$ are the distinct roots of $f_s(1, X)$. Thus the integer d is the number of distinct tangent lines of C.

We call the union $\bigcup_{j=1}^d C_j'$ of the disjoint algebroid curves C_j' the <u>proper quadratic transform</u> T(C) of C.

The quadratic transformation T operates also on the (X_1, X_2)-plane: it blows up the origin 0 into the line $X_1 = 0$ of the plane (X_1, X_2'). We call this line the <u>exceptional</u> curve of T and we denote it by E'; this curve contains the $\underline{d}$ origins $0_1', 0_2', \ldots, 0_d'$. We set $C_j'^* = C_j' \cup E'$, and we call the union $\bigcup_{j=1}^d C_j'^*$ of the d algebroid curves $C_j'^*$ the <u>total quadratic transform</u> $T\{C\}$ of C.

Let $p_1, p_2, \ldots, p_d$ be the distinct tangent lines of C, where we assume that p_j is associated with the root c_j of $f_s(1, X)$. We denote by C_j the union of those branches of C which have p_j as tangent line, and we call the algebroid curves $C_1, C_2, \ldots, C_d$ the <u>tangential components</u> of C. We have thus $C = \bigcup_{j=1}^d C_j$, and one sees easily that $C_j' = T(C)$, $C_j'^* = T\{C_j\}$, $j = 1, 2, \ldots, d$.

O. Zariski

3. Three (equivalent) Definitions of Equivalence of Plane Algebroid Curves.

We denote by $\gamma_1, \gamma_2, \ldots, \gamma_h$ the h branches of C. Let D be another algebroid curve, which has the same number of branches as C. Let $\delta_1, \delta_2, \ldots, \delta_h$ be the branches of D.

A (1;1) mapping π of the set of branches of C onto the set of branches of D will be said to be a tangentially stable pairing between the branches of C and D if the following is true: for any pair of branches $\gamma_\alpha, \gamma_\beta$ of C the corresponding branches $\pi(\gamma_\alpha), \pi(\gamma_\beta)$ have the same tangent if and only if $\gamma_\alpha, \gamma_\beta$ have the same tangent.

We shall use the notation: $\pi : (C) \rightarrow (D)$ to indicate a tangentially stable pairing between the branches of C and D.

It is clear that if there exists a tangentially stable pairing $\pi : (C) \rightarrow (D)$, then the number of distinct tangents of C is the same as the number of distinct tangents of D. If $p_1, p_2, \ldots, p_d$ are the tangents of C and $q_1, q_2, \ldots, q_d$ are the tangents of D, then, given $\pi : (C) \rightarrow (D)$, we have an induced (1,1) mapping of the set $\{p_1, p_2, \ldots, p_d\}$ onto the set $\{q_1, q_2, \ldots, q_d\}$. We shall assume that p_j and q_j correspond in this mapping.

It is well known that by a finite number $\sigma(C)$ (respectively $\sigma^*(C)$) of successive quadratic transformations it is possible to obtain a proper transform of C which has no singular points (respectively, a total transform of C which has only ordinary double points). This integer $\sigma(C)$, or the integer $\sigma^*(C)$, can be regarded as a measure of the complexity of the singularity of C. This fact is the basis of the various inductive definitions of equivalence $C \equiv D$ given below. Namely, in each of these definitions it is assumed that the meaning of the equivalence statements $C'_j \equiv D'_j$ and $C'^*_j \equiv D'^*_j$ is already known for $j = 1, 2, \ldots, d$ (as $\sigma(C'_j) < \sigma(C)$ and $\sigma^*(C'^*_j) < \sigma^*(C)$, etc.) and it is also stipulated that if C is a regular branch (respectively, if C has an ordinary double point) then $C \equiv D$ if and only also D is a regular branch (respectively, if also D has an ordinary double point).

<u>Definition 3.1. : A tangentially stable pairing</u> $\pi : (C) \to (D)$ <u>is an</u> (a) - <u>equivalence if the following conditions are satisfied:</u>

1) <u>If</u> $\delta_i = \pi(\gamma_i)$ $(i=1, 2, \ldots, h)$, <u>then</u> $m(\gamma_i) = m(\delta_i)$.

2) <u>The pairings</u> $\pi'_j : (C'_j) \to (D'_j)$ <u>induced by</u> π <u>are</u> (a) - <u>equivalences.</u>

It is understood that if $h = 1$ and C is a regular branch, in which case π is uniquely determined, then π <u>is</u> an (a) -equivalence if and only if condition 1) is satisfied, i.e., if and only if also D is a regular branch. This is also our understanding in Definition 3;2; given below.

In the above definition we have made use of the obvious fact that any tangentially stable pairing $\pi: (C) \to (D)$ induces a (1, 1) mapping π_j of the set of branches of any of the d tangential components C_j of C onto the set of branches of the <u>corresponding</u> tangential component D_j of D (the common tangent of the branches of C_j and the common tangent of the branches of D_j being <u>corresponding</u> tangents p_j and q_j). The mapping π_j induces a pairing π'_j between the branches of

C'_j and those of D'_j .

It should also be pointed out that according to the above definition, if a pairing $\pi:(C) \to (D)$ is an (a)-equivalence then it must be at any rate tangentially stable. Thus, condition 2) includes the tacit requirement that the π'_j be tangentially stable.

Let $\pi:(C) \to (D)$ be a tangentially stable pairing and let π'_j be defined as above. We extend π'_j to a (1, 1) mapping π'_j between the branches of C'^*_j $(=C'_j \cup E')$ and the branches of D'^*_j $(=D_j \cup E'_1)$, by setting $\pi'^*_j(E') = E'_1$; here E'_1 is the exceptional curve which corresponds to the origin of D .

<u>Definition 3.2:</u> <u>A tangentially stable pairing</u> $\pi:(C) \to (D)$ <u>is a</u> (b)-<u>equivalence if the following conditions are satisfied:</u>

1) <u>The pairings</u> $\pi'_j : (C'_j) \to (D'_j)$ <u>are</u> (b) -<u>equivalences.</u>
2) <u>The pairings</u> $\pi'^*_j:(C'^*_j)\to(D'^*_j)$ <u>are</u> (b-<u>equivalences)</u>. (j=1, 2, . . . , d) .

It shall be understood that if h = 2 and C has an ordinary double point, then π is a (b)-equivalence if and only if also D has an ordinary double point. Note that in this case there are two (1, 1) mappings of the pair of branches (γ_1, γ_2) of C onto the pair of branches (δ_1, δ_2) of D and that both are tangentially stable pairings, hence are (b)-equivalences according to our convention.

<u>Definition 3.3;</u> C <u>and</u> D <u>are formally equivalent if there exists a tangentially stable pairing</u> $\pi:(C) \to (D)$ <u>such that:</u>

1) C'_j <u>and</u> D'_j <u>are formally equivalent</u> $(j = 1, 2, \ldots, d)$.

2) C'^*_j <u>and</u> D'^*_j <u>are formally equivalent</u> $(j = 1, 2, \ldots, d)$.

It is understood here that if C is a regular branch (respectively, if C has an ordinary double point), then C and D are formally equivalent if and only if also D is a regular branch (respectively, if also D has an ordinary double point) .

It can be shown - and this is not entirely a trivial matter - that <u>these three definitions are all equivalent</u>; Definition 3.1 is the only one in which the multiplicities of the two curves C, D (and hence also, inductively, the multiplicities of the connected components of their successive quadratic transforms) are explicitly required to be the same [in view of condition 1) of that definition] . This definition is also the one which adheres most faithfully to the classical concept of equivalence which is due to Max Noether and which is based on the notion of multiple points of C which are "finitely near" the "proper" point 0 (the origin) and lie in the successive neighborhoods of order 1, 2 etc. of 0 (a pattern which is often referred to as the <u>composition</u> of the singular point 0 of C) .

In Definitions 3.2. and 3.3 the multiplicites of the branches of C or of D (and hence also the multiplicities of C and D themselves-) are not mentioned at all; these two definitions are therefore more subtle than Definition 3.1 (and, for that very reason, more useful for some applications, as we shall see later on) . On the other hand, in both Definitions 3.2. and 3.3, the inductive hypothesis includes not only the equivalence ((b)-equivalence in

Definition 3.2; formal equivalence in Definition 3.3) of the connected components C_j', D_j' of the proper transforms of C and D, but also the equivalence of the connected components $C_j'^*$, $D_j'^*$ of the total transforms of C and D. The difference between Definition 3.2 and 3.3 is this: a (b)-equivalence is by definition, a certain tangentially stable pairing $\pi:(C) \rightarrow (D)$, while formal equivalence is a certain condition on the pair $\{C.D.\}$ but is not a tangentially stable pairing $(C) \rightarrow (D)$. More precisely, while in Definition 3.3 we do assume the existence of a tangentially stable pairing $\pi:(C)\rightarrow(D)$, nothing is said about the nature of the pairings $\pi_j' : (C_j') \rightarrow (D_j')$ and $\pi_j'^* : (C_j'^*) \rightarrow (D_j'^*)$ induced by π. Definition 3.3 is the most subtle of the three definitions.

4. Outline of Proof of Equivalence

For any two algebroid curves C.D in the plane, free from common branches, we consider the intersection number (C, D). This number is zero if and only if the two curves have distinct origins.

Assuming that C and D have the same origin 0 and applying a quadratic transformation T with center 0, we have - in the notations of § 3 - the following well-known formula :

(1) $$(C, D) = m(C) \;\; m(D) + \sum_{\alpha=1}^{d} \sum_{\beta=1}^{\bar{d}} (C'_\alpha, D'_\beta) ,$$

where $C'_1, C'_2, \ldots, C'_d$ are the connected components of $T(C)$ and $D'_1, D'_2, \ldots, D'_{\bar{d}}$ are the connected components of $T(D)$. In particular, then, $(C, D) = m(C)m(D)$ if and only if C and D have no common tangents. If, furthermore, E' is the exceptional curve into which 0 is blown up by T, then we also have

(2) $$\sum_{\alpha=1}^{d} (C'_\alpha, E') = m(C) .$$

The two formulas (1) and (2) are all that is technically needed for the proof, which consists of several steps.

(A) One first proves the following lemma :

Lemma 4.1. Let $\pi : (C) \to (D)$ be an (a)-equivalence, let Γ be a branch through the origin of C, let Δ be a branch through the origin of D and assume that Γ and Δ are regular branches, that Γ is not a branch of C and that Δ is not a branch of D. Assume furthermore that

(3) $$(\gamma_i, \Gamma) = (\delta_i, \Delta) , \quad i = 1, 2, \ldots, h ,$$

where $\gamma_1, \gamma_2, \ldots, \gamma_h$ are the branches of C; $\delta_1, \delta_2, \ldots, \delta_h$ are the branches of D, and where $\delta_i = \pi(\gamma_i)$. Extend the mapping π to a mapping ρ of the set $\{\gamma_1, \gamma_2, \ldots, \gamma_h, \Gamma\}$ onto the set $\{\delta_1, \delta_2, \ldots, \delta_h, \Delta\}$ by setting $\rho(\Gamma) =$. Then $\rho : (C \cup \Gamma) \to (D \cup \Delta)$ is an (a)-equivalence.

The proof of this lemma is by induction on the character

$\sigma^*(C \cup \Gamma)$. If $\sigma^*(C \cup \Gamma) = 0$, then C is a regular branch, C and Γ have distinct tangents, $(C, \Gamma) = 1$, whence also $(D, \Delta) = 1$, by (3). Thus also $\sigma^*(D \cup \Delta) = 1$, and the assertion of the lemma is then trivial.

It follows directly from (3) that ρ is a tangentially stable pairing which satisfies condition 1) of (a)-equivalence. In the proof that ρ also satisfies condition 2) of Definition 3.1 one applies a quadratic transformation to C and D and one considers separately two cases accordingly as the tangent of Γ is or is not a tangent of C.

(B) The next step is the following

Lemma 4.2. *If* $\pi : (C) \to (D)$ *is an* (a)-*equivalence and* $\pi(\gamma_i) = \delta_i$ $(i = 1, 2, \ldots, h)$, *then*

$$(\gamma_i, \gamma_j) = (\delta_i, \delta_j), \text{ for all } i \neq j .$$

The proof is by induction on the numerical character $\sigma(C)$ and is a straightforward consequence of formula (1).

(C) Using Lemmas 4.1 and 4.2, one proves

Theorem 4.3. *If* $\pi : (C) \to (D)$ *is an (a)-equivalence, then it is also a (b)-equivalence; and conversely.*

The proof is by induction on $\sigma^*(C)$ and is rather straightforward. At this stage we drop the terms (a)-equivalence and (b)--equivalence and speak simply of equivalence pairings $\pi : (C) \to (D)$, and we say that C and D are *equivalent* if there exists an equivalence pairing $\pi : (C) \to (D)$. The proof that equivalence of C and D implies the formal equivalence of C and D is immediate, and is by induction on $\sigma^*(C)$. For if $\pi:(C)\to D$ is an equivalence pairing, then π, being a (b)-equivalence pairings $\pi'_j:(C'_j)\longrightarrow(D'_j)$ and $\pi'^*_j:(C'^*_j)\to(D'^*_j)$ (j = 1, 2, ..., h). Therefore, by induction, C'_j and D'_j are formally equivalent and so are C'^*_j and D'^*_j. Hence C and D are formally equivalent.

More difficult is the proof of the converse, i.e., that formal equivalence implies equivalence. The key to the proof is the following lemma:

Lemma 4.4. Let C and D have the same number h of branches. Let $\gamma_1, \gamma_2, \dots \gamma_m$ be certain m of the h branches of $C (m \leqq h)$ and assume that these m branches are regular and have distinct tangents. Similarly, let $\delta_1, \delta_2, \dots$ ..., δ_m be m of the m of the h branches of D, also regular and having distinct tangents. Let

$$C = \Gamma \cup \gamma_1 \cup \gamma_2 \cup \dots \cup \gamma_m ,$$
$$D = \Delta \cup \delta_1 \cup \delta_2 \cup \dots \cup \delta_m ,$$

where Γ, therefore, is the union of the remaining branches of C, and similarly for Δ and D. Assume that Γ and Δ are equivalent and that for each $\alpha = 1, 2, \dots, m$ also $\Gamma \cup \gamma_1 \cup \gamma_2 \cup \dots \cup \gamma_\alpha$ and $\Delta \cup \delta_1 \cup \delta_2 \cup \dots \cup \delta_\alpha$ are equivalent. Then there exists an equivalence pairing $\pi:(C)\longrightarrow(D)$, such that $\pi(\gamma_\alpha) = \delta_\alpha$, for $\alpha = 1, 2, \dots, m$.

The proof of this lemma is divided into several cases :

Case 1. m = 1, and the tangent of γ_1 is not a tangent of

Γ. This case is straightforward: the equivalence of Γ and Δ implies that Γ and Δ have the same number of tangent lines; similarly, $\Gamma \cup \gamma_1$ and $\Delta \cup \delta_1$ have the same number of tangent lines. Since the number of tangent lines of $\Gamma \cup \gamma_1$ is one greater than the number of tangent lines of Γ, it follows that the tangent of δ_1 is not a tangent of Δ. From the definition of (a)-equivalence it follows at once that given an equivalence pairing $\rho : (\Gamma) \to (\Delta)$, the extended pairing $\pi : (C) \to (D)$ defined by $\pi(\gamma_1) = \delta_1$, is an (a)-equivalence.

<u>Case 2</u>. $m = 1$ <u>and the tangent of</u> γ_1 <u>is a tangent of</u> Γ <u>(whence also the tangent of</u> δ_1 <u>is a tangent of</u> Δ).

In this case, one first achieves a reduction to the case in which Γ (and therefore also C) has only one tangent line. Passing to the total quadratic transforms of C and D (a transition which adds a new regular branch to both total transforms, namely the exceptional curve created by the quadratic transformation), one finds oneself in the case in which $m = 2$, while $\sigma^*(C)$ has decreased. The result now follows by induction on $\sigma^*(C)$.

<u>Case 3</u>. $m > 1$. In this case the result follows easily by applying the case $m = 1$ to each of the tangential components $C_1, C_2, \ldots, C_m$ of C which are determined by the m tangents of $\gamma_1, \gamma_2, \ldots, \gamma_m$.

The proof that formal equivalence implies equivalence is now as follows :

In the notation of Definition 3. 3 (§ 3), we have, by hypothesis, that C_j' and D_j' are formally equivalent and that also $C_j' \cup E'$ and $D_j' \cup E'$ are formally equivalent (we assume here, without loss of generality, that C and D have the same origin 0). By induction on σ (C), we can conclude that C_j' and

D_j' are equivalent and that also $C_j' \cup E'$ and $D_j' \cup E'$ are equivalent. Therefore, by the case m = 1 of Lemma 4.4 , there exists an equivalence $\pi_j':(C_j' \cup E') \to (D_j' \cup E')$ such that $\pi_j'(E') = E'$. But that implies that C and D are equivalent (namely, (b)-equivalent).

Note: We have used here only the case m=1 of Lemma 4.4. However, the proof of the lemma for m=1 required that the lemma be proved also in the case m = 2 (see Case 2 above). So Lemma 4.4. is needed only for m=1, 2.

For later applications we ahall also need the following equivalence criterion:

Proposition 4.5. Assume that C and D have the same number h of branches and that a pairing π between the branches γ_i of C and the branches δ_i of D has the following two properties:

1) Corresponding branches γ_i δ_i are equivalent (i=1, 2, ..., h) .

2) For all pairs (i, j) , i= j, we have $(\gamma_i, \gamma_j) = (\delta_i, \delta_j)$. Then π is an equivalence.

Proof. By 1) we have $m(\gamma_i) = m(\delta_i)$, i = 1, 2, ..., h. Then 2) implies that π is tangentially stable. Let $C_1, C_2, \ldots, C_d$ be the tangential components of C and let $D_1, D_2, \ldots, D_d$ be the corresponding tangential components of D (relative to π). Apply a quadratic transformation to both C and D, and let C_j' and D_j' be the proper transforms of C_j and D_j. Let π_j' be the pairing between the branches of C_j' and D_j' induced by π . If γ' is any branch of C_j' and $\delta' = \pi_j'(\gamma')$ is the corresponding branch of D_j', then we have, by 1), that γ' and δ' are equivalent. If γ_1', γ_2' are any two branches of C_j' and δ_1', δ_2' are the corresponding branches of D_j', we have that γ_1', γ_2' are the quadratic transform of two branches γ_1, γ_2 of C_j, while δ_1', δ_2'

are the quadratic transforms of the corresponding branches δ_1, δ_2 of D_j.

We have $(\gamma_1', \gamma_2') = (\gamma_1, \gamma_2) - m(\gamma_1)m(\gamma_2)$, $(\delta_1', \delta_2') = (\delta_1, \delta_2) - m(\delta_1)m(\delta_2)$. Hence $(\gamma_1', \gamma_2') = (\delta_1', \delta_2')$. The proposition now follows by induction on the numerical character $\sigma(C)$.

In connection with Proposition 4.5, it is well to recall at this time the well known criterion of equivalence of algebroid branches γ, δ, in terms of the characteristic exponents of the Puiseux expansions of γ and δ.

Let (x, y) be a minimal basis of the local domain $\mathfrak{O}$ of γ and assume that x is a transversal parameter. Then, if n denotes the multiplicity of γ $(n > 1)$, we can expand y into a power series in $x^{1/n}$;

$$(4) \qquad y = \sum_{i}^{\infty} = n\, a_i x^{i/n} ,$$

where the highest common divisor of all the numerators i such that $a_i \neq 0$ is prime to n. (In the present case we actually have $a_i = 0$ for $i < n$.) The characteristic exponents.

$$(5) \qquad \beta_1/n < \beta_2/n < \ldots < \beta_g/n, g \geqq 1$$

of the Puiseux expansion (4) are defined by the following conditions:

a) $n > (n, \beta_1) > (n, \beta_1, \beta_2) > \ldots > (n, \beta_1, \beta_2, \ldots, \beta_g) = 1$;

b) $a_{\beta_j} \neq 0$, $j = 1, 2, \ldots, g$;

c) Each β_j is minimum (more precisely: if $\beta_1, \beta_2, \ldots, \beta_{j-1}$ have already been defined $(j \geqq 1)$ and if $(n, \beta_1, \beta_2, \ldots, \beta_{j-1}) > 1$, then β_j is the smallest of all the integers q such that $a_q \neq 0$ and $(n, \beta_1, \beta_2, \ldots, \beta_{j-1}) >$

$$> (n, \beta_1, \beta_2, \ldots, \beta_{j-1}, q)) .$$

Since n is the least common denominator of the rational numbers β_j/n, the characteristic exponents determine the integer n, hence the multiplicity of γ. Let

$$\bar{y} = \sum_{i=n}^{\infty} \bar{a}_i \bar{x}^{i/\bar{n}} , \tag{6}$$

be the Puiseux expansion of the branch δ, where $\bar{x}$ is also a transversal parameter of δ (whence $\bar{n} = m(\delta)$), and let

$$\bar{\beta}_1/\bar{n} < \bar{\beta}_2/\bar{n} < \ldots \bar{\beta}_{\bar{g}}/\bar{n} \tag{7}$$

be the characteristic exponents of (6). We have the following well-known result :

Proposition 4.6. <u>The branches γ and δ are equivalent if and only if the Puiseux expansion (4) and (6) have the same characteristic exponents.</u>

[It is understood that it is implicitly assumed in this proposition that x and $\bar{x}$ are transversal parameters; or -equivalently - that $\beta_1 > n$ and $\bar{\beta}_1 > \bar{n}$].

A fast proof of this proposition, avoiding the complications arising from relating the compositinn of the singularity of γ to the expansions of the characteristic exponents into continued franctions, can be obtained by using the following <u>inversion formula</u> (see Abhyankar [2] ; Zariski [9] , p.996) :

Let $a_m x^{m/n}$ <u>be the leading term of</u> (4) $(m \geqq n)$, <u>and let</u>

$$\beta'_1/m < \beta'_2/m < \ldots < \beta'_{g'}/m$$

be the characteristic exponents of the Puiseux expansion $x = \dot{x}(y^{1/m})$.
Then :

1) If $n < m < \beta_1$ (i.e., if $m > n$ and is an integral multiple of n), then $g' = g + 1$, $\beta'_1 = n$, $\beta'_{\nu+1} = \beta_\nu + n - m$ ($\nu = 1, 2, \ldots, g$).

2) In all other cases (i.e., if., $m = n$ or $m = \beta_1$), we have $g' = g$ and $\beta' = \beta_\nu + n - m$ ($\nu = 1, 2, \ldots, g$).

We now apply to γ and δ the quadratic transformations $y' = y/x$ and $\bar{y}' = \bar{y}/\bar{x}$ respectively. Without loss of generality, we may assume that the leading terms of (4) and (6) are $a_{\beta_1} x^{\beta_1/n}$ and $a_{\bar{\beta}_1} \bar{x}^{\bar{\beta}_1/\bar{n}}$ The transformed branches γ', δ' will be centered at the origin, and will be given by the expansions:

$$\gamma' : \quad y' = \sum_{i=\beta_1}^{\infty} a_i x^{(i-n)/n} \tag{8}$$

$$\delta' : \quad \bar{y}' = \sum_{i=\bar{\beta}_i}^{\infty} \bar{a}_i \bar{x}^{(i-\bar{n})/\bar{n}} . \tag{9}$$

I. Assume $\gamma \equiv \delta$. Then $n = \bar{n}$ and $\gamma' \equiv \delta'$.

(a) If $\beta_1 > 2n$, then $m(\gamma') = m(\gamma)$ $(=n)$, whence also $m(\delta') = m(\delta)$, and thus $\bar{\beta}_1 > 2n$. In this case, x and $\bar{x}$ are still transversal parameters for γ' and δ', and using induction on $\sigma(\gamma)$, we conclude that the characteristic exponents of (8) and (9) are the same, i.e., $(\beta_i - n)/n = (\bar{\beta}_i - n)/n$, all i, and the proposition is proved.

(b) If $\beta_1 < 2n$, we have $m(\gamma') = \beta_1 - n < n$. Therefore also $m(\delta') < m(\delta)$, showing that $\bar{\beta}_1 = \beta_1$. This time, y' and $\bar{y}'$ are transversal parameters. The induction hypothesis implies now that the characteristic exponents of the two Puiseux expansions $x = x(y'^{1/(\beta_1 - n)})$, $\bar{x} = \bar{x}(\bar{y}'^{1/(\beta_1 - n)})$ are the same. The proposition now follows easily from the inversion formula.

II. Assume that (4) and (6) have the same characteristic exponents.

We have now again $n = \bar{n}$, whence $m(\gamma) = m(\delta)$. As in Case I, there are again the two cases (a) and (b) to be considered. In case (a), the equality of the characteristic exponents of (8) and (9) implies without further ado (and by induction hypothesis) that $\gamma' \equiv \delta'$, and hence $\gamma \equiv \delta$. In case (b), the equality of the characteristic exponents of (8) and (9) implies, by the inversion formula, the equality of the characteristic exponents of the expansions $x = x(y'^{1/(\beta_1 - n)})$, $\bar{x} = \bar{x}(\bar{y}'^{1/(\beta_1 - n)})$, and hence we find again that $\gamma' \equiv \delta'$.

Corollary. The characteristic exponents of (4) (with x-transversal) are independent of the choice of the transversal parameter x.

Apply Proposition 4.6 to the case $\gamma = \delta$. We can therefore speak of the characteristic exponents of the branch γ.

5. On Some Numerical Characters of C'

Given a plane algebroid curve C and given a parameter x such that x belongs to some minimal basis $\{x, y\}$ of the maximal ideal $\mathfrak{m}$ of the local ring $\mathfrak{o}$ of C, we shall associate with the pair $\{C, x\}$ an integer $M_x(C)$ defined as follows:

Let y be any element of $\mathfrak{m}$ such that $\{x, y\}$ is a basis of $\mathfrak{m}$ and let $f(X, Y) = 0$ be the equation of C relative to this basis. Then $f(X, Y)$ is regular in Y, and we may assume then that f is a monic polynomial in Y and that $f(0, Y)$ is a power Y^n of Y (these conditions determine f uniquely). In this sense we can speak of the Y-discriminant $\Delta^Y f$ of f; it will be a power series in X. Up to a unit factor in $k[[X]]$, this power series is independent of the choice of y, and hence depends only on the pair $\{C, x\}$. The integer $M_x(C)$ is the highest power of X which divides $\Delta^Y f$.

Assume C irreducible. Then the quotient field of the local domain $\mathfrak{o}$ is a field $k((C))$ of formal power series in one indeterminate $\{u\}$. We denote by v the natural valuation of $k((C))$. Assume that both elements x and y of the basis $\{x, y\}$ of $\mathfrak{m}$ are parameters; equivalently: $f(X, Y)$ is also regular in X. Then also the integer $M_y(C)$ is defined. It is easy to see that

(1) $$M_x(C) = v\left(\frac{\partial f}{\partial y}\right), \quad M_y(C) = v\left(\frac{\partial f}{\partial x}\right).$$

We have the relations $\frac{\partial f}{\partial x}\,dx + \frac{\partial f}{\partial y}\,dy = 0$, $v\left(\frac{dx}{du}\right) = v(x) - 1$, $v\left(\frac{dy}{du}\right) = v(y) - 1$. Hence, by (1), we find that

(2) $$M_x(C) - v(x) = M_y(C) - v(y).$$

Thus the integer $M_x(C) - v(x)$ is therefore a numerical character of the irreducible curve C. Using the well-known relation in the integral closure $\bar{\mathfrak{o}}$ of the local domain $\mathfrak{o}$:

$$(f_y) = \mathfrak{C}\,\mathfrak{D}_x ,$$

where $\mathfrak{C}$ is the conductor of $\mathfrak{o}$ in $\bar{\mathfrak{o}}$ and $\mathfrak{D}_x$ is the different of $\bar{\mathfrak{o}}$ in $\mathfrak{o}$, and recalling that (since k is of characteristic zero) $\mathfrak{D}_x = \bar{\mathfrak{m}}^{e-1}$, where $e = v(x)$ and $\mathfrak{m}$ is the maximal ideal of $\bar{\mathfrak{o}}$, it follows that

$$\mathfrak{C} = \bar{\mathfrak{m}}^{M_x(C)-v(x)+1}$$

Whether C is or is not irreducible, it is easy to see that any transversal parameter x belongs to some minimal basis. Hence the integer $M_x(C)$ is defined for all transversal parameters x of o . In the case of an irreducible C , what we have just said abov shows that (a) the integer $M_x(C)$ is the same for all transversal parameters, for if x is transversal then $v(x) = m(C)$; it is also obvious that (b) if $M_{x'}$ is defined and x' is not transversal then $M_x(C) < M_{x'}(C)$. We can show easily that the last two assertions are also valid if C is not irreducible. For, let $\gamma_1, \gamma_2, \ldots, \gamma_h$ be the branches of C. The local ring $\mathfrak{o}_i$ of γ_i is a homomorphic image of $\mathfrak{o}$, and if x_i, y_i are the images of x, y in $\mathfrak{o}_i$, then (x_i, y_i) is a basis of the maximal ideal of $\mathfrak{o}_i$, and x_i is a parameter of $\mathfrak{o}_i$. From the expression of $\Delta^Y f$ in terms of the roots $Y_1, Y_2, \ldots, Y_n$ of $f(X, Y)$ follows at once that

$$(3) \qquad M_x(C) = \sum_{i=1}^{h} M_{x_i}(\gamma_i) + 2 \sum_{i<j} (\gamma_i, \gamma_j) .$$

From this relation, the above assertions (a), (b) follow also if C is reducible. We denote by $\Delta(C)$ the integer $M_x(C)$ x - a transversal parameter of o. It follows from (3) that

$$(4) \qquad \Delta(C) = \sum_{i=1}^{h} \Delta(\gamma_i) + 2 \sum_{i<j} (\gamma_i, \gamma_j) .$$

Proposition 5.1. If $C \equiv D$, then $\Delta(C) = \Delta(D)$.

Proof. I view of (4) and of Lemma 4.2, it is sufficient to consider the case in which C and D are irreducible. We shall assume that C and D have been embedded in the same (X, Y)-plane, that they have the same origin $X = Y = 0$, and that the line $X = 0$ is neither a tangent of C nor a tangent of D. Let $\mathfrak{o}, \mathfrak{o}_1$ be the local domains of C and D, let $\{x, y\}$ be a minimal basis of the maximal ideal of $\mathfrak{o}$, and let $\{x_1, y_1\}$ be a minimal basis of the maximal ideal of $\mathfrak{o}_1$. We assume that both x and x_1 are transversal parameters. We may assume that neither y nor y_1 is transversal. If we then set $y' = y/x$, $y_1' = y_1/x_1$, and call C', D' the quadratic transforms of C and D, we see that $\{x, y'\}$ is a basis of the maximal ideal of the local ring of C', and that similarly $\{x_1, y_1'\}$ is a basis of the maximal ideal of the local ring of D'. We set $s = m(C) = m(D)$. One finds at once that

$$M_x(C) = s(s-1) + M_x(C') ,$$

$$M_{x_1}(C) = s(s-1) + M_{x_1}(D') ,$$

So we have only to show that $M_x(C') = M_{x_1}(D')$.

If $m(C') = m(C)$, then also $m(D') = m(D)$ (since $C' \equiv D'$), x, x_1 are still transversal parameters for C' and D', $M_x(C') = \Delta(C')$, $M_{x_1}(D') = \Delta(D')$, and the proposition follows by induction on the integer $\Delta(C)$. Assume, however, that $m(C') < m(C)$, and that consequently x is not a transversal parameter for C' (nor, then, is x_1 transversal for D'). The local equation of the exceptional curve E', in the (X, Y')-plane, is $X = 0$, and this line is tangent to C', since $(E', C') = m(C)$ (formula (2), § 4) and $m(C) > m(C')$. Similarly, E' is tangent to D'. Therefore the line $Y' = 0$ is neither a tangent of C' nor a tangent of D'. By induction on $\Delta(C)$, we may therefore assume that $M_{y'}(C') = M_{y_1'}(D')$, since these two integers are equal respectively to $\Delta(C')$ and $\Delta(D')$. We have relations similar to (2):

$$M_x(C') - v(x) = M_{y'}(C') - v(y');$$

$$M_{x_1}(D') - v_1(x_1) = M_{y_1'}(D') - v_1(y_1'),$$

where v_1 is the natural valuation of the field $k((D))$. Now, $v(y') = m(C')$, $v_1(y_1') = m(D')$, thus $v(y') = v_1(y_1')$. Similarly $v(x) = v_1(x_1)$ $(= m(C) = m(D))$. The proposition now follows from $M_{y'}(C') = M_{y_1'}(D')$.

Proposition 5.1 can be strengthened as follows:

<u>Proposition 5.2.</u> <u>Let</u> (x, y) <u>be a minimal basis of the maximal ideal of the local ring of</u> C, <u>let</u> (x_1, y_1) <u>be a minimal basis of the maximal ideal of the local ring of</u> D. <u>Consider the associated embeddings of</u> C <u>and</u> D <u>in the</u> (X, Y)-<u>plane and denote by</u> ℓ <u>the line</u> $X = 0$. <u>Assume that</u> $C \equiv D$ <u>and</u>

that there exists an equivalence pairing $\pi : (C) \to (D)$, such that $(\ell, \gamma) = (\ell, \delta)$ for any pair of corresponding branches γ and $\delta = \pi(\gamma)$. Then $M_x(C) = M_{x_1}(D)$.

Proof. By (3) and by Lemma 4.2, it is sufficient to consider the case in which C and D are irreducible. If ℓ is not tangent to C, then our assertion coincides with Proposition 5.1. Assume that ℓ is tangent to C, and hence also to D (since $m(C) = m(D)$). By using relations similar to (2) and the notations of the proof of Proposition 5.1, we find that

$$M_x(C) = \Delta(C) + (\ell, C) - m(C),$$

$$M_{x_1}(D) = \Delta(D) + (\ell, D) - m(D),$$

and the proposition is proved.

6. Equisingular Analytic Families of Algebroid Curves: A Discriminant Criterion

We consider an analytic family

(1) $$\mathfrak{F}: f(X, Y; t) = 0$$

of plane algebroid curves, depending on a parameter t and having the same origin $X = Y = 0$ (whence $f(0, 0; t)$ is identically zero; without any change in what follows, except for more cumbersome notations, we could have assumed that the family $\mathfrak{F}$ depends on a finite number of parameters $t_1, t_2, \ldots, t_\rho$). Here f is a formal power series in X, Y, t, with coefficients in our algebraically closed ground field k (of characteristic zero). We make the following assumptions: f has no multiple factors in $k[[X, Y; t]]$, and if we set

(2) $$f_o(X, Y) = f(X, Y; 0),$$

then f_o is not identically zero and has no multiple factors in $k[[X, Y]]$. We denote by K the algebraic closure of the field $k((t))$, and we regard (1) as an equation of an algebroid curve C_t, defined over K, and with origin $X = Y = 0$; C_t is then the general member of the family $\mathfrak{F}$. We denote by C_o the special member of $\mathfrak{F}$, defined by

(3) $$C_o : f_o(X, Y) = 0,$$

where f_o is defined in (2). Thus C_o is an algebroid curve defined over k; but we shall regard it also as defined over K. We say that $\mathfrak{F}$ is an equisingular family, if the two algebroid curves C_t and C_o (both having their origin at $X = Y = 0$ and both being defined over K) are equivalent (i.e., have equivalent singularities). Our object in this section is to give a discriminant criterion for $\mathfrak{F}$ to be equisingular and to outline the proof of that criterion. This criterion is of basic importance for our theory of equisingularity in codimension 1 (see § 7).

We shall first introduce some notations and make some preliminary remarks. We denote by $\mathfrak{o}_t$ the (complete) local ring of C_t (over K), by $\mathfrak{o}$ the (complete) local ring of C_o (over k), and by $\mathfrak{O}$ the local ring $k[[X, Y; t]]/(f) = k[[x, y; t]]$, where x, y are the f-residues of X, Y (we identify t with its f-residue). Then (x, y) is a basis of the maximal ideal $\mathfrak{m}_t$ of

$\mathfrak{o}_t$. We have $\mathfrak{o} = \mathfrak{O}/\mathfrak{O}t = k[[\xi,\eta]]$, where ξ, η are the $\mathfrak{O}t$-residues of x, y. Then (ξ, η) is a basis of the maximal ideal $\mathfrak{m}$ of $\mathfrak{o}$. We assume that X does not divide $f_o(X, Y)$, or-equivalently - that ξ is a parameter of $\mathfrak{o}$. Then, a <u>fortiori</u>, X does not divide $f(X, Y;t)$, and x is a parameter of $\mathfrak{o}_t$. We denote by $\mathfrak{M}$ the maximal ideal (x, y, t) of $\mathfrak{O}$.

We shall <u>not</u> allow the minimal basis (x, y) of $\mathfrak{m}_t$ to range over the entire set of minimal bases of $\mathfrak{m}_t$. Rather, only such other bases (x', y') of $\mathfrak{m}_t$ will be allowed in <u>which the element</u> x', y' <u>belong</u> to $\mathfrak{O}$; or-equivalently - we must have $x' = Ax + By$, $y' = Cx + Dy$, where $A, B, C, D \in \mathfrak{O}$ and $AD - BC$ is a unit in $\mathfrak{O}$. If ξ', η' denote the $\mathfrak{O}t$-residues of x', y', then (ξ', η') is a basis of $\mathfrak{m}$. The further conditinn that ξ' be a parameter of $\mathfrak{o}$ (and hence that x' be a parameter, of $\mathfrak{o}_t$) remains in force.

Any basis (x, y) of $\mathfrak{m}_t$, fixed according to the above specifications, will determine an associated embedding (1) of C_t in the (X, Y)-plane, with $f(X, Y;t) \in k[[X, Y;t]]$, and an associated embedding (3) of C_o, where f_o is defined in (2) and where X does not divide f_o.

We consider the decomposition of f into its irreducible factors in $k[[X, Y;t]]$:

$$(4) \qquad f = \prod_{i=1}^{h} f_i(X, Y;t) .$$

Since we have assumed that $f(0, 0;t) = 0$, we must have $f_i(0, 0;t) = 0$ for at least one of the factors f_i. Let s_i denote the degree of the initial form of $f_i(X, Y;t)$, where we now regard f_i <u>as a power series</u> in X, Y (over K). <u>A priori</u>, we may have

$s_i = 0$ for <u>some</u> i. If, say, $s_1 = 0$, then $f_1(X, Y;t)$ is a unit in $K[[X, Y]]$ and the factor f_1 could be deleted from the equation $f = 0$ without affecting C_t. We have

$$(5) \qquad f_o = \prod_{i=1}^{h} f_i(X, Y; 0) = \prod_{i=1}^{h} f_{i,o}(X, Y) .$$

The degree $s_{i,o}$ of the initial form of $f_i(X, Y;0)$ is $\geqq s_i$. Since $m(c_t) = \sum s_i$, $m(C_o) = \sum s_{i,o}$, it follows that <u>if</u> $C_t \equiv C_o$ <u>then necessarily</u> $s_i = s_{io}$ <u>for</u> $i = 1, 2, \ldots, h$. Now, each $f_i(X, Y;t)$ is irreducible, hence a non - unit in $k[[X, Y]];t$. Therefore $f_i(0, 0; 0) = 0$, showing that $s_{io} > 0$. Hence, <u>a necessary condition for the family $\mathfrak{F}$ to be equisingular is that s_i be positive for all</u> $i = 1, 2, \ldots, h$, i.e., <u>that</u> $f_i(0, 0;t)$ <u>be identically zero for each i.</u> However, we shall not impose this condition explicitly in our criterion; it will follow automatically from that criterion.

A final preliminary remark is the following :

If ℓ denotes the line $X = 0$ then $(\ell, C_t) \leqq (\ell, C_o)$. If, then, $m(C_t) = m(C_o)$ (in particular, if $C_t \equiv C_o$), and if we choose the embedding (1) of C_t in such a way that ℓ is <u>not</u> tangent to C_o (equivalently: ξ is a transversal parameter of $\mathfrak{O}$), then necessarily ℓ will also not be a tangent of C_t. (and hence $(\ell, C_t) = (\ell, C_o)$). Thus, <u>if</u> $C_t \equiv C_o$, <u>there always exist embeddings</u> (1) of C_t, <u>such that the intersection multiplicities</u> (ℓ, C_t) <u>and</u> (ℓ, C_o) <u>are equal</u> (where ℓ is the line $X = 0$); in particular, this will be so if ℓ is not tangent to C_o (always under the assumption that $C_t \equiv C_o$).

Since f is regular in Y, we may assume that f is a monic polynomial in Y (with coefficients in $k[[X;t]]$), such that $f(0, Y; 0)$ is a power of Y, say $f(0, Y;0) = Y^n$.

(these conditions determine f uniquely). We denote by $D(X;t)$ the Y-discriminant of f; this will be a power series in X and t. We now state our criterion :

Theorem 6.1 *The following two conditions are equivalent:*

(A) C_t C_o *and* $(\ell, C_t) = (\ell, C_o)$ *(where* ℓ *is the line* $X = 0$) .

(B) $D(X,t)$ *is of the form* $X^M \varepsilon(X,t)$, *where* $\varepsilon(X,t)$ *is a is a unit in* $k[[X,t]]$ (i.e., $\varepsilon(0,0) \neq 0$) .

Note that the condition $(\ell, C_t) = (\ell, C_o)$ signifies that $f(0,Y;t) = Y^n$, or -equivalently - that $f_i(0,0;t)$ is identically zero for each of the h irreducible factors in (4) ; thus this property is a consequence of the criterion (B), as we have anticipated in our preceding remarks. We outline the proof of this theorem.

a) The starting point is the following well-known result (see for instance S Abhyankar [1] , Th. 4, p.585, and O. Zariski [7], Theorem 5.p.527) :

If $f(X,Y;t)$ *is irreducible in* $k[[X,Y;t]]$ *and is a monic polynomial in* Y, *say* $f(0,Y;0) = Y^n$, *and if condition* (B) *of the theorem is satisfied, then the roots* $Y_1, Y_2, \ldots, Y_n$ of f *(in the algebraic closure of* $k((X,t)))$ *belong to the power series ring* $k[[X^{1/n};t]]$.

Therefore, the n (conjugate) roots are given by Puiseux expansions

$$(6) \qquad Y_q = \sum_{\nu=0}^{\infty} u_\nu(t) \omega^{q\nu} X^{\nu/n}, \quad (q = 1, 2, \ldots, n)$$

where the $u_\nu(t)$ are in $k[[t]]$, $u_0(0) = 0$, and ω is a primitive n^{th} root of unity. We have thus :

$$f(X, Y;t) = \prod_{q=1}^{n} (Y-Y_q),$$

and

(7) $$f(0, Y;t) = [Y - u_o(t)]^n .$$

b) The conclusion (7), i.e., the property that $f(0. Y;t)$ is the nth power of $Y-u_o(t)$, where $u_o(t) \in k[[t]]$, remains true if $f(X, Y;t)$ is reducible [the other assumptions in a) remaining in force]

For, with reference to the factorization (4) of f, it is clear that condition (B) implies that this same condition is satisfied for each irreducible factor f_i of f, i.e., we have

(B_i) $$\Delta^Y f_i = X^{M_i} \varepsilon_i(X;t)$$

where $\varepsilon_i(X;t) \in k[[X; t]]$ and $\varepsilon_i(0; 0) \neq 0$. Hence, by a), it follows that if $f_i(0, Y;0) = Y^{n_i}$ then the n_i roots $Y_q^{(i)}$ of f_i are given the Puiseux expansion

$$Y_q^{(i)} = \sum_{\nu=0}^{\infty} u_{\nu,i}(t)\omega_i^{q\nu} X^{\nu/n_i}, \quad (q = 1, 2, \ldots, n_i)$$

where ω_i is a primitive n_i^{th} proof of unity, and that $f_i(0, Y;t) = [Y - u_{o,i}(t)]^{n_i}$. Our assertion in b) is to the effect that $u_{o,1}(t) = u_{o,2}(t) = \ldots = u_{o,h}(t) \; (=u_o(t))$. For assuming the contrary : $u_{o,i}(t) \neq u_{o,j}(t)$, for some pair of indices i, j in the set $\{1, 2, \ldots, h\}$, it would follow that if $Y_q^{(i)}$ and $Y_\rho^{(j)}$ denote any root of f_i and f_j respectively, then the difference $Y_q^{(i)} - Y_\rho^{(j)}$ does not divide X^M in $k[[X^{1/n_1 n_2 \cdots n_h}; t]]$, contrary to condition (B) .

Since we have assumed initially that $f(0,0;t) = 0$, it follows that in our present case we must have $u_o(t) = 0$, i.e. , <u>we must have as a consequence of</u> (B) .

$$(8) \qquad f_i(0, Y; t) = Y^{n_i}, \quad i = 1, 2, \ldots, h .$$

c) <u>If we assume</u> (B) <u>and if</u> f <u>is irreducible in</u> $k[[X, Y;t]]$, <u>then</u> $f(X, Y; 0)$ <u>is irreducible in</u> $k[[X, Y]]$.

For, in the notations of a), we must have that the roots $Y_{1,o}, Y_{2,o}, \ldots, Y_{n,o}$ of $f(X, Y;0)$ are

$$(9) \qquad Y_{q,o} = \sum_{\nu=o}^{\infty} u_\nu (o)\omega^{q\nu} X^{\nu/n}, \qquad q = 1, 2, \ldots, n ,$$

and thus the irreducibility of $f(X, Y; 0)$ follows from our original assumption that $f(X, Y;0)$ has no multiple factors.

d) From here on , until further notice, we assume that condition (B) is satisfied. As a consequence of (B) C_t has precisely h branches $\gamma_{t,1}\gamma_{t,2}, \ldots, \gamma_{t,h}$, (and not less) , where $\gamma_{t,i}$ is defined by the irreducible equation $f_i(X, Y; t) = 0$. If we now apply c) we conclude that the specialization γ_i of $\gamma_{t,i}$, i.e., <u>the algebroid curve</u> $f_{i,o}(X, Y) = 0$, where $f_{i,o}(X, Y) = f_i(X, Y;0)$, <u>is also irreducible</u> and that therefore C_o also has exactly h branches. We have therefore a particular (1, 1) mapping of the set of branches of C_t onto the set of branches of C_o , namely the mapping π defined by $\pi(\gamma_{t,i}) = \gamma_i$ $(i = 1, 2, \ldots, h)$. We call π the <u>specialization pairing</u> between the branches of C_t

and those of C_o

e) We have $m(\gamma_{t,i}) = m(\gamma_i)$, $i = 1, 2, \ldots, h$.

For the proof one may assume that C_t and C_o are irreducible. Let $u_m(t)$ be the first coefficient $u_\nu(t)$ in (6) which is different from zero (necessarily $m > 0$ since $u_o(t) = 0$). Then $m(C_t) = \min.\ m, n$. If $m \geqq n$, then it follows from (9) that $m(C_t) = m(C_o) = n$. Assume $m < n$, whence $m(C_t) = m$. Then the proof is completed by observing that we must have in (9) $u_m(0) \neq 0$. For, assuming the contrary, i.e., assuming that $u_m(t)$ is divisible by t, we would find that the difference of the roots

$$Y_1 - Y_n = u_m(t)\ \{\omega^m - 1\}\ X^{m/n} + \ldots$$

does not divide X^M in $k[[X^{1/n}\ ; t]]$, in contradiction with (B).

f) <u>The specialization pairing</u> π <u>is tangentially stable.</u> We have by (3) , § 5 :

$$M = \sum_{i=1}^{h} M_i + 2 \sum\sum_{i<j} (\gamma_{t,i}, \gamma_{t,j}) \tag{10}$$

and also

$$M = \sum_{i=1}^{h} M_i + 2 \sum\sum_{i<j} (\gamma_i, \gamma_j), \tag{11}$$

where the M_i are defined in b), (B_i).

Furthermore, we have $(\gamma_{t,i}, \gamma_{t,j}) \leqq (\gamma_i, \gamma_j)$, since π is a specialization pairing . This implies, by (10) and (11) , that

(12) $$(\gamma_{t,i}, \gamma_{t,j}) = (\gamma_i, \gamma_j) ,$$

and this, in view of e), implies that π is tangentially stable.

g) We have already shown in b) [see (8)] that (B) implies the second part of condition (A), namely $(\ell, C_t) = (\ell, C_o)$ $(= n = \sum_{i=1}^{h} n_i)$, where ℓ is the line $X = 0$, We now prove that (B) implies (A) by showing that $C_t \equiv C_o$. More precisely, we shall show that the <u>specialization pairing</u> $\pi : (C_t) \to (C_o)$ is an <u>equivalence.</u> By (12), and in view of Proposition 4.5, it is sufficient to show that corresponding branches $\gamma_{t,i}, \gamma_i$ of C_t and C_o are equivalent. <u>We may therefore assume that C_t and C_o are irreducible.</u> We show now that it is also permissible to assume that <u>the line</u> $\ell : X = 0$ is not a tangent of C_o [and, hence, <u>a fortiori,</u> is not a tangent of C_t, since we know already, by e) that $m(C_t) = m(C_o)$]. For, assume that $X = 0$ is a tangent of C_t (and hence also of C_o). Then the line $Y = 0$ is not a tangent of C_o (since C_o is irreducible, it has only one tangent line). Let v and v_o be the natural valuations of $K((C_t))$ and $k((C_o))$ respectively and let $s = m(C_t) = m(C_o)$. Then $v(x) = v_o(\xi) = n$, while $v_o(y) = v_o(\eta) = s < n$. We have, by (2), § 5 :

$$M_y(C_t) = M_x(C_t) - n + s ,$$

$$M_\eta(C_o) = M_\xi(C_o) - n + s .$$

Since $M_x(C_t) = M_\xi(C_o) = M$, by (B), it follows that $M_y(C) = M_\eta(C_o) =$ = (say) N. This shows that $\Delta^X f$ is of the form $Y^N \varepsilon_1(Y;t)$, where $\varepsilon_1(Y;t) \in k[[Y;t]]$ and $\varepsilon_1(0;0) \neq 0$. Thus (B) is still satisfied when we interchange the roles of X and Y. [In forming the discriminant $\Delta^{X}{}_{f}$ we must, of course multiply first f be a suitable unit factor in k [[X, Y;t]] so as to make f to be a monic polynomial in X (of degree s)].

We assume therefore that the line ℓ : X = 0 is not tangent to C_o (whence ξ and x are transversal parameters of $\mathfrak{o}$ and $\mathfrak{o}_t$ respectively) . Without loss of generality we may assume that the line Y = 0 is tangent to C_o (replace Y by Y + cX, where c is a suitable element of k). We set $\overline{Y} = Y/X$ and

$$f(X, X\overline{Y};\, t) = X^{s}\overline{f}(X, \overline{Y};t) .$$

Then the quadratic transforms $\overline{C}_t = TC_t)$ and $\overline{C}_o = T(C_o)$ of C_t and C_o are given respectively by the equations

$$\overline{C}_t : \overline{f}(X, \overline{Y};t) = 0 ,$$

$$\overline{C}_o : \overline{f}_o(X, \overline{Y}) = 0, \quad \text{where } \overline{f}_o(X, \overline{Y}) = \overline{f}(X, \overline{Y};0) .$$

The origin of $\overline{C}_o$ is $X = \overline{Y} = 0$. The $\overline{Y}$ - discriminant $\Delta^{\overline{Y}}{}_{\overline{f}}$ is given by $^{Y}{}_{f}/X^{s(s-1)}$, and hence

$$(13) \qquad \Delta^{\overline{Y}}{}_{\overline{f}} = x^{M-s(s-1)}\,\varepsilon\,(X, t) ,\ \varepsilon(0;0) \neq 0 .$$

Thus (B) is satisfied for the algebroid curve $\overline{C}_t$. We have $\overline{f}(0, 0;0) = \overline{f}_o(0, 0) = 0$, and hence, by b) , we may therefore assert that the polynomial $\overline{f}_o(X, \overline{Y};t)$ (in $\overline{Y}$) has a root of the form $\overline{Y} = \overline{u}_o(t)$, where $\overline{u}_o(t) \in k\,[[t]]$ and $\overline{u}_o(0) = 0$.

We replace $\overline{Y}$ by $\overline{Y} - \overline{u}_o(t)$. This has no effect on $\overline{C}_o$

(since $\bar{u}_o(0) = 0$), but for the new $\bar{C}_t$ (or for the new embedding of $\bar{C}_t$) we will have $\bar{f}(0, 0;t) = 0$. By induction on $\Delta^Y f$, we may therefore assume that $\bar{C}_t \equiv \bar{C}_o$. Hence $C_t \equiv C_o$, as asserted.

h) We now complete the proof of the theorem by showing that (A) implies (B). Since $C_t \equiv C_o$, the two curves C_t and C_o must have the same number of branches. This implies that the number h of irreducible factors of $f(X, Y;t)$ [see (4)] must be the number of branches of C_t [since $f_i(0, 0, ;0) = 0$ for $i = 1, 2, \dots, h$, and since $f(X, Y;0)$ has, by assumption, no multiple factors, the number of branches of C_o is at least h) and that $f_{i,o}(X, Y)$ ($= f_i(X, Y;0)$) is irreducible in $k[[X, Y,]]$. This yields a specialization pairing π between the h branches $\gamma_{t,i} : f_i(X, Y;t) = 0$ of C_t and the h branches $\gamma_i : f_{i,o}(X, Y) = 0$ of C_o. We have, of course, $(\ell, \gamma_{t,i}) \leqq (\ell, \gamma_i)$, for $i = 1, 2, \dots, h$. However, since $(\ell, C_t) = \sum_{i=1}^{h} (\ell, \gamma_{t,i})$ and $(\ell, C_o) = \sum_{i=1}^{h} (\ell, \gamma_i)$, it follows from our assumption $(\ell, C_t) = (\ell, C_o)$ [included in (A)] that $(\ell, \gamma_{t,i}) = (\ell, \gamma_i)$. If, then, we can prove that the specialization pairing π is an equivalence, then (B) will follow from Proposition 5.2. Now, we prove that the assumption $C_t \equiv C_o$ implies that π is an equivalence. In this proof we may assume that the line $X = 0$ is not tangent to C_o (and therefore not tangent to C_t). Then $M_x(C_t) = \Delta(C_t)$ and $M_\xi(C_o) = M(C_o)$. Since $C_t \equiv C_o$, it follows from Proposition 5.1 that $M_x(C_t) = M_\xi(C_o) =$ (say) M. This implies (B), and (B) in its turn has been shown already to imply [see g) above] that π is an equivalence.

7. Equisingularity in Codimension 1.

As a matter of interpretation, we can look at the equation $f(X, Y; t) = 0$ in two ways: (1) as in § 6, as an equation defining an analytic family $\mathfrak{F}$ of plane algebroid curves in the (X, Y) plane, with the common origin $X = Y = 0$, and depending on a parameter t; (2) or as the equation of an algebroid surface F in the (X, Y, t)-three space, with origin at $X = Y = t = 0$ and containing the line $W : X = Y = 0$. Then the curves of the family $\mathfrak{F}$ can be looked upon as sections $t =$ const. of F by planes normal to the line W. To our way of thinking, the property of the family $\mathfrak{F}$ of being equisingular corresponds, in the second way of looking at the equation $f = 0$, to the property of the surface F of being equisingular at the origin $X = Y = t = 0$ along the line W, which is our way of expressing an intuitive statement that the singularity which F has at the special point $X = Y = t = 0$ of the line W is "not worse" than (or is "of the same type" as) the singularity of F at the generic point (0, 0, t) of W. We shall now develop this idea through precise definitions and we will add to Theorem 6.1 a number of other criteria of equisingularity, so as to give an adequate picture of our theory of equisingularity in codimension 1. As was pointed out in § 6, the assumption that we have only one parameter t was of no importance whatsoever. We shall, in fact, assume that we have r - 1 parameters $t_1, t_2, \dots, t_{r-1}$ or - equivalently - that we are dealing with an algebroid hypersurface $f(X, Y; t_1, t_2, \dots, t_{r-1}) = 0$ in an affine (r + 1)-space.

Let

(1) $$V_r : f(X_1, X_2, \ldots, X_r, X_{r+1}) = 0$$

be an algebroid hypersurface in an affine (r+1)-space A_{r+1}, with the origin Q at $X_1 = X_2 = \ldots = X_{r+1} = 0$. Here f is a formal power series in the X's, with coefficients in our ground field k. We assume that f has no multiple factors in $k[[X_1, X_2, \ldots, X_{r+1}]]$, that $f(0, 0, \ldots, 0) = 0$ and and that the origin Q is a singular point of V. We consider an irreducible analytic subvariety W of V, or codimension 1 on V (hence of dimension r-1), passing through Q and having at Q a simple point. We may then choose the local (analytic) coordinates X_i in the neighborhood of Q in $\mathbb{A}_{r+1}$ in such a way that W is defined by

(2) $$W : X_r = X_{r+1} = 0 .$$

We denote by R the general point $(t_1, t_2, \ldots, t_{r-1}, 0, 0)$ of W, where the t's are analytically independent parameters.

We consider sections C_o of V with algebroid surfaces L_2 passing through $\mathcal{O}$, having at $\mathcal{O}$ a simple point, and transversal to W at $\mathcal{O}$. Any such L_2 can be defined by equations of the form

(3) $$L_2 : X_i = A_{i1}(X)X_r + A_{i2}(X) \ X_{r+1}, \qquad i = 1, 2, \ldots, r-1 ,$$

where the A_{ij} are formal power series in $X_1, X_2, \ldots, X_{r+1}$, with coefficients in 1. It is clear that when L_2 is given then we can choose the local coordinates X_i in $\mathbb{A}_{r+1}$ in such a way that L_2 is given by the equations

(4) $$L_2 : X_i = 0 \quad , \qquad i = 1, 2, \ldots, r-1 \ .$$

We require, furthermore, that $L_2 \not\subset V$. Thus, if L_2 is given by the equations (4), then we require that $f(0, 0, \ldots, 0\ X_r, X_{r+1})$ not be identically zero. We call then the section C_o of V with L_2 a W-transversal section of V, at Q. A section C_o may have multiple components [$f(0, 0, \ldots 0, X_r, X_{r+1})$ may have multiple factors], and is thus an algebroid cycle of dimension 1 (a plane algebroid curve, if C_o has no multiple components).

On the contrary, up to an analytical isomorphism there is only one W-transversal section C_t of V, at the general point R of W (see Zariski [8], p. 980); if we set $\mathfrak{o}$ = local ring of V:

$$\mathfrak{o} = k[[x_1, x_2, \ldots, x_{r+1}]] = k[[X_1, X_2, \ldots, X_{r+1}]]\ (f) \ ,$$

and if $\mathfrak{p}$ denotes the prime ideal of W in $\mathfrak{o}$:

$$\mathfrak{p} = \mathfrak{o}x_r + \mathfrak{o}x_{r+1} \ ,$$

then C_t is a plane algebroid curve, whose local ring is equal to (up to isomorphism) the completion of the local ring $\mathfrak{o}_{\mathfrak{p}}$. A typical C_t is the curve defined by

(5) $$C_t : f(t_1, t_2, \ldots, t_{r-1}; X_r, X_{r+1}) = 0 \ .$$

Note that $k((t_1, t_2, \ldots, t_{r-1}))$ is k-isomorphic to the residue field of the local ring $\mathfrak{o}_{\mathfrak{p}}$ and that C_t is defined over this field, and hence also over the algebraic closure K of $k((t_1, t_2, \ldots, t_{r-1}))$.

Definition 7.1. V is said to be equisingular at Q, along W, if there exists a W-transversal section C_o of W at Q, such that 1) C_o is a curve (i.e., has no multiple components) and such that 2) $C_t \equiv C_o$ (it is implicit in this definition that dim $W = r-1$ and Q is a simple point of W).

Definition 7.2. A singular point Q of V is said to be a singularity of dimensionality type 1 (for V) if there exists an irreducible subvariety W of codimension 1, containing Q, such that V is equisingular at Q along W.

The elements $x_1, x_2, \ldots, x_r, x_{r+1}$ form a minimal basis of the maximal ideal of $\mathfrak{O}$, and they determine the embedding (1) of V. The elements $x_1, x_2, \ldots, x_r$ are parameters of $\mathfrak{O}$ if and only if f is regular in X_{r+1}, and when that is we may assume that f is a monic polynomial in X_{r+1} and that

$$f(0, 0, \ldots, 0, X_{r+1}) = X_{r+1}^n . \tag{6}$$

If s is the multiplicity of the singular point Q of V, then $n \geqq s$, with equality if and only if the parameters $x_1, x_2, \ldots x_r$ are transversal. Clearly, if $x_1, x_2, \ldots, x_r$ are parameters and if W is defined by (2), then the section C_o of V defined by (4) is transversal to W. Conversely, if the (4) define a transversal section, then there exists a linear combination $x_r' = cx_r + dx_{r+1}$, with $c, d \in k$, such that $x_1, x_2, \ldots, x_{r-1}, x_r'$ are parameters of $\mathfrak{O}$. From now on we deal only with parameters $x_1, x_2, \ldots, x_r$ of $\mathfrak{O}$ which can be extended to a basis $\{x_1, x_2, \ldots, x_r, x_{r+1}\}$ of the maximal ideal of $\mathfrak{O}$

We assume that $x_1, x_2, \ldots, x_r$ are parameters of $\mathfrak{O}$, we denote by $D(X_1, X_2, \ldots, X_r)$ the X_{r+1}-discriminant of the monic polynomial f [of degree n in X_{r+1}; see (6)], by D_o the product of the distinct irreducible factors of D, and by $\Delta_{(x)}$ the algebroid (r-1)-dimensional bypersurface in A_r defined by the equation $D_o(X_1, X_2, \ldots, X_r) = 0$. We call $\Delta_{(x)}$ the critical variety of V associated with the parameters $x_1, x_2, \ldots, x_r$. This variety depends only on $x_1, x_2, \ldots, x_r$ (and is independent of the choice of x_{r+1}).

Definition 7.3. A system of parameters $x_1, x_2, \ldots, x_r$ of $\mathfrak{O}$ is said to be equisingular at Q if the associated critical variety $\Delta_{(x)}$ has a simple point at its origin $X_1 = X_2 = \ldots = X_r = 0$.

The criterion of equisingular analytic families of algebroid curves, derived in § 6, leads (after some additional considerations) to the following criterion of equisingularity in codimension 1 (Zariski [8], Theorems 4.4, 4.5 and 5.2):

Theorem 7.4. The following conditions are equivalent :

(a) The point Q is a singularity of dimensionality type 1.

(b) There exist equisingular systems of parameters at Q

(c) Every system of transversal parameters at Q is equisingular.

Incidental results such as the following can also be proved :

1) If Q is a singular point of dimensionality type 1; then the singular locus S of V has, locally at Q, codimension

1 and a simple point (but not conversely). Thus, the singular points of the singular locus S and also the singular points of V which do not lie on (r-1)-dimensional components of S are not of dimensionality type 1.

2) If the parameters $x_1, x_2, \ldots, x_r$ form an equisingular system, then the principal ideal generated by $D_o(x_1, x_2, \ldots, x_r)$ in $\mathfrak{O}$ is primary, of codimension 1, and the irreducible subvariety W defined by the associated prime ideal of $\mathfrak{O}\, D_o$ has a simple point at 0, with V equisingular at Q along W. More explicitly: assuming, as we may, that $D_o(x_1, x_2, \ldots, x_r) = x_r$, we can extend $(x_1, x_2, \ldots, x_r)$ to a minimal basis $(x_1, x_2, \ldots, x_r, x_{r+1})$ of the maximal ideal of $\mathfrak{O}$, such that the ideal $\mathfrak{O}x_r + \mathfrak{O}x_{r+1}$ is the prime ideal of the primary ideal $\mathfrak{O}x_r$.

3) In defining W-transversal sections C_o of V, at Q, we have not only imposed conditions on the surface L_2 whose intersection with V is C_o [see equation (4) of L_2], but we have also required that $L_2 \not\subset V$. In Definition 7.1 of equisingularity we have required, furthermore, that C_o be a curve (i.e., free from multiple components). Now, from part c) of Theorem 7.4, it follows easily that if V is equisingular at Q, along W, then: 3a) for every surface L_2 defined as in (3), we have $L_2 \not\subset V$; 3b) every W-transversal section of V is a curve (i.e., is free from multiple components); 3c) all W-transversal sections of V are equivalent algebroid curves.

4) If V is an algebraic hypersurface then the set of singular points of V which are not of dimensionality type 1 is an algebraic subvariety of V of dimension $\leq r-2$. To see this, let, say, V be projective. We project V onto a projective r-space $\mathbb{P}_r$ using as center of projection various points 0 of $\mathbb{P}_{r+1} - V$, we call π_0 the associated regular map $V \to \mathbb{P}$, we denote by Δ_0 the critical hypersurface (in $\mathbb{P}_r$) of the projection π_0, by S_0 the singular locus of Δ_0, and then we conclude, by (c), that the set of singular points in question is the intersection $\bigcap \pi_0^{-1}\{S_0\}$, as 0 ranges over $\mathbb{P}_{r+1} - V$.

In particular, if V is an algebraic surface, then we call exceptional singularities of V the singular points of V which are not of dimensionality type 1. These points are finite in number, and they include, among others, a) the isolated singular points of V (i.e., the singular points which do not belong to singular curves of V) and b) the singular points of the singular curve of V (if the singular locus of V has dimension 1).

5) The singularities of dimensionality type 1 behave very nicely with respect to monoidal transformations $T: V' \to V$ centered at irreducible (r-1)-dimensional components W of the singular locus S of V. Namely, with any such W we can associate an equivalence class $\{C_W\}$ of plane algebroid curves, where $\{C_W\}$ is the class which contains the W-transversal section C_t of V at the general point R of

W. By Definition 7.1 and by 3), if V is equisingular along W at a point Q of W, then all the W-transversal sections of V at Q belong to the class $\{C_W\}$. Now, let $\underline{d}$ be the number of distinct tangent lines of a member C of the class $\{C_W\}$, let C' be the quadratic transform of C and let $C_1', C_2', \ldots, C_d'$ be the connected component of C', which correspond to the tangential components $C_1, C_2, \ldots, C_d$ of C. The equivalence classes $\{C_1'\}, \{C_2'\}, \ldots, \{C_d'\}$ depend only on the equivalence class $\{C_W\}$ [see § 3, Definition 1 of (a)-equivalence]. This being said, the following can be shown :

If Q is any point of W such that V is equisingular at Q along W, then the full inverse image $T^{-1}\{Q\}$ of Q on V' consists precisely of d points $Q_1', Q_2', \ldots, Q_d'$. If W' is the full inverse image $T^{-1}\{W\}$ of W on V', then one and only one irreducible component W_i' of W' passes through Q_i' $(i = 1, 2, \ldots, d$; we may have, of course $W_i' = W_j'$ for $i = j)$; we have $\dim W_i' = r-1$, and V' is equisingular at Q_i', along W_i'. Furthermore, for a suitable labeling of the points Q_i', the equivalence class $\{C_{W_i}\}$ coincides with the above equivalence class $\{C_i'\}$ $(i = 1, 2, \ldots, d)$.

If follows that if V is equisingular at a point Q or W, then the singularity of V at Q is completely determined by the type of singularity of the W-transversal sections of V at Q, i.e., by the equivalence class $\{C_W\}$, and that the singularity of V at Q is resolved by a sequence of consecutive monoidal transformations, with $(r-1)$-dimensional centers, which runs altogether parallel to the sequence of successive quadratic transformations which are required to desingularize the equivalence class $\{C_W\}$. Since we have just seen that T doesn not blow up such a point Q, the desingularization of the singular point Q of V amounts, in the present case, to the normalization of V at Q.

In particular, if V is a surface which has no exceptional singularities, then the normalization of V is non-singular and can be obtained from V by monoidal transformations centered at curves.

6) If k is the field of complex numbers, there is the following topological property of equisingularity in codimension 1, which is due to Whitney [6] :

If V is equisingular at a point A of W, and if C_o is a W-transversal section of V at Q, then V, as an embedded variety in the complex affine $\mathbb{A}_{r+1}$, is topologically equivalent (more precisely: isotopic), locally at Q, to the direct product $C_o \times W$.

8. A Saturation-Theoretic Criterion of Equivalence of Algebroid Curves.

As an introduction to our new concept of saturated local rings (to be defined later on in this section and in § 9) and to a special type of equisingularity in codimension > 1, based on that concept we shall give here a new criterion of equivalence of algebroid plane curves. For simplicity we shall illustrate our idea on the case of irreducible curves.

Let, then, C be an algebroid branch in the (X, Y)-plane. We go back to § 4 and we consider there the Puiseux expansion (4) of C. We assume that x is a transversal parameter of the local domain $\mathfrak{o}$ of C, hence that

$$y = \sum_{i=m}^{\infty} a_i x^{i/n} \qquad m \geqq n > 1 \tag{1}$$

Here n = m (C) (=multiplicity of the origin 0 of C). We consider the characteristic exponents of the branch C [§4, (5)]:

$$\beta_1/n < \beta_2/n < \ldots < \beta_g/n, \qquad g \geqq 1 .$$

Let F be the field of fractions of $\mathfrak{o}$:

$$\mathfrak{o} = k[[x]][y],$$
$$F = k((t)), \quad t = x^{1/n} .$$

Let G be the Galois group of F/K, where K = k((x)). This is a cyclic group of order n. Let G_i be the subgroup of G which has order equal to the highest common divisor

$$(n, \beta_1, \beta_2, \ldots, \beta_i) ,$$

so that we have now the strictly descending chain

$$G = G_o > G_1 > G_2 > \ldots > G_g = 1 . \tag{2}$$

The following is easily verified:

"$\tau \varepsilon G_{i-1}, \tau \notin G_i$" $\Longrightarrow$ "$y^{\tau} - y = a_{\beta_i}(\omega_{\tau} - 1)x^{\beta_i/n}$ + terms of higher degree".

Here ω_{τ} is a nth root of unity, <u>different from</u> 1. Hence if $\underline{v}$ is the natural valuation of $F(v(t) = 1,\ v(x) = n)$ then

(3) $$v(y^{\tau} - y) = \beta_i,\ \tau \varepsilon G_{i-1}, \tau \notin G_i \quad (i = 1, 2, \ldots, g) .$$

This yields another characterization of the characteristic exponents β_i/n of the branch C :

$$\{\beta_1/n, \beta_2/n, \ldots, \beta_g/n\} = \{v(y^{\tau} - y)/n \mid \tau \varepsilon G, \tau \neq 1\} , \ n = m(C).$$

Now suppose that we have another irreducible plane algebroid branch C' , given by

(1') $$y' = \sum_{m'}^{+\infty} a_i' x'^{i/n'} , \quad m' \geqq n' > 1 .$$

Let $\mathfrak{O}'$ be the complete local ring of C', and F' the field of fractions of $\mathfrak{O}'$. Let us assume that C and C' have equivalent singularities. Then, in the first place, we must have $n' = n$, since these are the multiplicities of the origin 0 of C and C' . Therefore, if we identify $F = k((t))$ with $F' = k(t'))$ in such a way that $T' = t$ ($t = x^{1/n}$, $t' = x'^{1/n'}$) , then we have identified x with x' . Let us make this identification. So now

(1 ') $$y' = \sum_{m'}^{+\infty} a_i' x^{i/n} , \quad m' \geqq n > 1 .$$

In the second place, the characteristic exponents of (1) and (1') must be the same. Therefore the chain (2) of subgroups of the Galois group G is the same for C and C', and hence by (3) we have

$$(4) \qquad v(y^{\tau}-y) = v(y'^{\tau}-y'), \quad \text{for all} \quad \tau \in G, \ \tau \neq 1 .$$

Let us set $\bar{\mathfrak{o}} = k[[t]]$ (=integral closure of both $\mathfrak{o}$ and $\mathfrak{o}'$ in F). Then, by (4):

$$(5) \qquad (y'^{\tau}-y')/(y^{\tau}-y) \in \mathfrak{o} \quad \text{and} \quad (y^{\tau}-y)/(y'^{\tau}-y') \in \mathfrak{o} , \quad \text{for all} \quad \tau \in G, \ \tau \neq 1 .$$

Let us consider the following ring contained between $\mathfrak{o}$ and $\bar{\mathfrak{o}}$:

$$(6) \qquad \tilde{\mathfrak{o}}_x = \{\xi \in \bar{\mathfrak{o}} \mid (\xi^{\tau}-\xi)/(y^{\tau}-y) \in \bar{\mathfrak{o}}, \ \text{all} \ \tau \in G, \tau \neq 1 .$$

It is immediate that $\tilde{\mathfrak{o}}_x$ is indeed a ring, hence a complete local ring, of dimension 1 (since both $\mathfrak{o}$ and $\bar{\mathfrak{o}}$ are complete local rings, of dimension 1). The ring $\tilde{\mathfrak{o}}_x$ can be characterized by the following 3 properties:

S_1) $\mathfrak{o} \subset \tilde{\mathfrak{o}}_x \subset \bar{\mathfrak{o}}$.

S_2) If $\zeta \in \tilde{\mathfrak{o}}_x$, $\eta \in \bar{\mathfrak{o}}$ and if $(\eta^{\tau}-\eta)/(\zeta^{\tau}-\zeta)\ \bar{\mathfrak{o}}$ for all $\tau \in G$[if for some τ, $\zeta^{\tau} = \zeta$, this last assumption should read as follows: $\eta^{\tau} = \eta$], then also $\eta \in \mathfrak{o}_x$.

S_3) $\tilde{\mathfrak{o}}_x$ is the smallest ring satisfying conditions S_1) and S_2).

Similarly, we set

$$(7) \qquad \tilde{\mathfrak{o}}'_x = \{\xi \in \bar{\mathfrak{o}} \mid (\xi^{\tau}-\xi)/(y'^{\tau}-y') \in \bar{\mathfrak{o}}, \quad \text{all} \ \tau \in G, \tau \neq 1 .$$

We now give the following definition :

Definiton 8.1 Let $\mathfrak{o}$ be the local ring of an irreducible algebroid branch C (not necessarily a plane branch), let x be a parameter of $\mathfrak{o}$, let $\bar{\mathfrak{o}}$ be the integral closure of $\mathfrak{o}$ in the quotient field F of $\mathfrak{o}$ and let G be the Galois group of F/K, where $K = k((x))$. We say that $\mathfrak{o}$ is saturated with respect to x if $\mathfrak{o}$ satisfies condition S_2.

Proposition-Definition 8.2 If $\mathfrak{o}$ and x are as above, there exists a unique ring $\tilde{\mathfrak{o}}_x$ satisfying conditions S_1, S_2, S_3. This ring is called the saturation of $\mathfrak{o}$ with respect to x.

We now go back to the inclusions (5) which hold if C and C' have equivalent singularities. These inclusions imply that $y' \in \tilde{\mathfrak{o}}_x$ and $y \in \tilde{\mathfrak{o}}'_x$, i.e.,

$$\mathfrak{o}' \subset \tilde{\mathfrak{o}}_x \quad \text{and} \quad \mathfrak{o} \subset \tilde{\mathfrak{o}}'_x ,$$

whence

$$\tilde{\mathfrak{o}}_x = \tilde{\mathfrak{o}}'_x . \tag{8}$$

We can express this result, as follows (we drop the identification $F = F'$):

Theorem 8.3 If $C \equiv C'$, then given any transversal parameter x' of $\mathfrak{o}'$, there exists an isomorphism $\varphi : \tilde{\mathfrak{o}}_x \xrightarrow{\sim} \tilde{\mathfrak{o}}'_{x'}$ such that $\varphi(x) = x'$, and conversely.

The truth of the converse can be seen as follows: If φ exists, then we must have in the first place $n=n'$, for in the respective natural valuations v, v' of F and F' we have $v(x) = n$, $v'(x')=n'$. Next, we can identify $\tilde{\mathfrak{o}}_x$ with $\tilde{\mathfrak{o}}'_{x'}$ and x with x'.

So now $\mathfrak{o}$ and $\mathfrak{o}'$ have the same field of fraction $F(=k((x^{1/n})))$, and, of course, the same integral closure $\bar{\mathfrak{o}}$ $(=k[[x^{1/n}]])$. Furthermore, we have $y' \in \mathfrak{o}' \subset \tilde{\mathfrak{o}}'_x = \tilde{\mathfrak{o}}_x$ and similarly $y \in \tilde{\mathfrak{o}}'_x$. Therefore, we have the inclusions (5), or-what is the same - the equalities (4), and these imply the equality of the characteristic exponents.

In the special case $C = C'$, Theorem 8.3 tells us that if x and x' are any two transversal parameters of $\mathfrak{o}$, then there exists an isomorphism $\varphi : \tilde{\mathfrak{o}}_x \xrightarrow{\sim} \tilde{\mathfrak{o}}_{x'}$ such that $\varphi(x) = x'$. The following is a far from trivial result and is not covered by what is known from the classical theory :

Theorem 8.4. _If_ x, x' _are any two transversal parameters of_ $\mathfrak{o}$, _then_ $\tilde{\mathfrak{o}}_x = \tilde{\mathfrak{o}}_{x'}$.

The local ring $\tilde{\mathfrak{o}}_x$, x-transversal, which is thus independent of x, shall be denoted by $\tilde{\mathfrak{o}}$ and c-lled _the saturation of_ $\mathfrak{o}$.

The saturation $\tilde{\mathfrak{o}}$ of $\mathfrak{o}$ is the local ring of some irreducible algebroid curve $\tilde{C}$ lying in an affine space $\mathbb{A}_N$ of some high dimension N (N can be computed explicitly in terms of the characteristic exponents). By Theorem 8.3 we have $C \equiv C'$ if and only if $\tilde{C} = \tilde{C}'$. Since $\mathfrak{o} \subset \tilde{\mathfrak{o}}$, C is a projection of $\tilde{C}$. Thus, _all irreducible plane algebroid branches in the same equivalence class are projections of one and the same algebroid branch_ $\tilde{C}$ _in some_ S_N, and - what is important - it can be shown that the _generic plane projection of_ $\tilde{C}$ _belongs to that same equivalence class._

We add without proof certain propositions on saturation.

If $\mathfrak{o}$ is the local ring of any irreducible (not necessarily plane) algebroid curve, we say that $\mathfrak{o}$ is _saturated_ if $\mathfrak{o}$ is saturated with respect to some parameter x of $\mathfrak{o}$. Then we have the following propositions :

Proposition 8.5. If $\mathfrak{o}$ is saturated, it is also saturated with respect to any transversal parameter.

Proposition 8.6 If $\mathfrak{o}$ is saturated and if x, x' are any two transversal parameters of $\mathfrak{o}$, then there exists an automorphism of $\mathfrak{o}$ which sends x into x'.

Thus, the automorphisms of a saturated local ring $\mathfrak{o}$ are as abundantly many as they could possibly be.

We also observe that Theorem 8.4 continues to hold true for local rings $\mathfrak{o}$ of algebroid branches which are not plane curves, and thus we can speak of the saturation $\tilde{\mathfrak{o}}$ of any such ring $\mathfrak{o}$, meaning by that the saturation of $\mathfrak{o}$ with respect to any transversal parameter. It can be shown that $\tilde{\mathfrak{o}}$ is the smallest saturated ring between $\mathfrak{o}$ and $\bar{\mathfrak{o}}$. The saturation of $\mathfrak{o}$ with respect to a parameter which is not transversal may very well depend on the choice of the parameter. It can be proved that various saturations of $\mathfrak{o}$ have one thing in common: they have the same multiplicity as $\mathfrak{o}$. We know that the family of these rings has a lower bound: it is the saturation of $\mathfrak{o}$ (i.e.), the saturation with respect to a transversal parameter of $\mathfrak{o}$). Now, if $n = e(\mathfrak{o})$, the largest local ring between $\mathfrak{o}$ and $\bar{\mathfrak{o}}$ which has multiplicity n is the ring $k + \bar{\mathfrak{m}}^n$, where $\bar{\mathfrak{m}}$ is the maximal ideal of $\bar{\mathfrak{o}}$. It can be shown that this is a saturation of $\mathfrak{o}$ with respect of some parameter x; it is sufficient to take x in $\mathfrak{m}$ so that $v(x)$ and n are relatively prime.

9. Saturation of Local Rings of Higher Dimension. Equisaturated Families of Hypersurfaces.

In the general theory of saturation we deal with an algebroid variety V of a given dimension r, embedded in some affine space. While in the case of dimension 1 we have restricted ourselves, for simplicity, to irreducible curves, in the general case I shall not assume that V is irreducible. However, I will assume that V is equidimensional. The ground field k is still assumed to be algebraically closed and of characteristic zero. The local ring $\mathfrak{O}$ of V may now have zero-divisors, but will have no nilpotent elements (we are dealing with algebroid varieties, or, if you wish, with reduced local formal schemes) We denote by F the total ring of quotients of $\mathfrak{O}$. Then F is a direct sum of fields :

(1) $$F = F_1 \oplus F_2 \oplus \ldots \oplus F_h ,$$

where h is the number of irreducible components of V. We have that k is a subfield of F. We write

(2) $$1 = \varepsilon_1 + \varepsilon_2 + \ldots + \varepsilon_h, \quad \varepsilon_i = \text{the identity of } F_i .$$

We fix any set of parameters $x_1, x_2, \ldots, x_r$ of $\mathfrak{O}$. It is well known that $\mathfrak{O}$ contains the ring of formal power series $R = k[[x_1, x_2, \ldots, x_r]]$. Our hypothesis that V is equidimensional implies that no element of R, different from zero, is a zero divisor in $\mathfrak{O}$. Hence F contains the field of fractions $K = k((x_1, x_2, \ldots, x_r))$ of R, and is, of course, a finite algebra over K. It is also known that $\mathfrak{O}$ is a finite R-module, hence is integral over R. We denote by $\bar{\mathfrak{O}}$ the integral closure of $\mathfrak{O}$ in F; this need not be a local ring, but is a semi-local ring, having exactly h maximal ideals. This, of course, has to do with the fact that normalization of V separates the irreducible branches of V .

I wish now to define the saturation $\tilde{\mathfrak{o}}_{(x)}$ of $\mathfrak{o}$ with respect to the parameters $x_1, x_2, \ldots, x_r$. In point of fact, the ring $\tilde{\mathfrak{o}}_{(x)}$ which will be thus defined, will depend not so much on the set of parameters x_i as on the field $K(=k(x_1, x_2, \ldots, x_r)))$ of formal meromorphic functions of these parameters. Therefore, it may be just as well to use the notation $\tilde{\mathfrak{o}}_K$. Now, in order to give this definition, it will be simpler to drop the assumption that $\mathfrak{o}$ is a local ring and to define $\tilde{\mathfrak{o}}_K$ for general commutative rings $\mathfrak{o}$ with identity, under the following conditions :

1. $\mathfrak{o}$ has no nilpotent elements :
2. The total ring of quotients F of $\mathfrak{o}$ is a finite direct sum of fields.
3. F contains a field K, and K contains the identity 1 of F.
4. Each field F_i (see (1)) is a finite separable extension of $K\epsilon_i$ (see (2) .
5. If $R = \mathfrak{o} \cap K$, then $\mathfrak{o}$ is integral over R .

All these conditions are automatically satisfied in the case in which $\mathfrak{o}$ is the local ring of the equidimensional algebroid variety V (in characteristic zero) and K is the field $k((x_1, x_2, \ldots, x_r))$.

We fix an algebraic closure Ω of K and we consider the set M of K-homomorphisms of F into Ω . The set M is finite, since F is a finite algebra over K. If $\psi \in M$, then

$\psi(\varepsilon_i) = 1$ for some i while $\psi(\varepsilon_j) = 0$ for $j \neq i$, and $\psi(F) = \psi(F_i)$ = isomorphic image of the field F_i. The composition of the various fields $\psi(F)$, $\psi \varepsilon M$, is a Galois extension F^* of K. In the special case in which $\mathfrak{o}$ is an integral domain we could have taken for Ω an algebraic closure of the field F, and F^* would then be the least Galois extension of K containing F.

Definition 9.1. If $\zeta, \eta \varepsilon F$, we say that η dominates ζ (with respect to K), in symbols: $\eta > \zeta$, if for any two homomorphisms ψ_1, ψ_2 in the set M we have that

$$[\psi_1(\eta) - \psi_2(\eta)] / [\psi_1(\zeta) - \psi_2(\zeta)] \text{ is integral over } R \text{ if } \psi_1(\zeta) \neq \psi_2(\zeta)$$

and

$$\psi_1(\eta) = \psi_2(\eta) \text{ if } \psi_1(\zeta) = \psi_2(\zeta).$$

Definition 9.2. We say that $\mathfrak{o}$ is saturated with respect to the field K if $\mathfrak{o}$ contains with any element ζ also every element η of $\bar{\mathfrak{o}}$ which dominates ζ.

It is immediate that any ring between $\mathfrak{o}$ and $\bar{\mathfrak{o}}$ satisfies conditions 1-5 and that the intersection of all the rings between $\mathfrak{o}$ and $\bar{\mathfrak{o}}$ which are saturated with respect to K (there exist such rings; for instance $\bar{\mathfrak{o}}$) is itself saturated with respect to K. We have therefore the following

Proposition-Definition 9.3. There exists a smallest ring between $\mathfrak{o}$ and $\bar{\mathfrak{o}}$ which is saturated with respect to K. This ring is called the saturation of $\mathfrak{o}$ with respect to K and is denoted by $\tilde{\mathfrak{o}}_K$.

A most important property of saturation is given by the following.

Proposition 9.4. If $R(= \mathfrak{O} \cap K)$ is integrally closed, then over every prime ideal $\mathfrak{p}$ of $\mathfrak{O}$ there lies one and only prime ideal $\tilde{\mathfrak{p}}$ of $\tilde{\mathfrak{O}}_K$, and the field of fractions of $\tilde{\mathfrak{O}}_K / \tilde{\mathfrak{p}}$ is a purely inseparable extension of the field of fraction of $\mathfrak{O}/\mathfrak{p}$.

In the Bourbaki terminology one would say that the natural morphism $\mathrm{Spec}(\tilde{\mathfrak{O}}_K) \longrightarrow \mathrm{Spec}(\mathfrak{O})$ is radicial.

Now, let us go back to the case in which $\mathfrak{O}$ is the local ring of an algebroid variety V. Since $\bar{\mathfrak{O}}$ is a semilocal ring and a finite module over $\mathfrak{O}$, every ring between $\mathfrak{O}$ and $\bar{\mathfrak{O}}$ is semi-local. However, Proposition 9.4, applied just to the maximal ideal $\mathfrak{m}$ of $\mathfrak{O}$ tells us that $\tilde{\mathfrak{O}}_K$ has only one maximal ideal, and is therefore again a local ring. It is therefore the local ring of an algebroid variety $\tilde{V}_K$ which dominates V and is dominated by the normalization $\bar{V}$ of V:

$$V \xleftarrow{f} \tilde{V}_K \xleftarrow{g} \bar{V}, \quad f \text{ and } g \text{ being morphisms.}$$

We call the variety $\tilde{V}_K$ the saturation on V with respect to K. Saturation, in general, falls short of normalization; in particular, it does not separate the branches of V. Furthermore, it can also be shown that the singular origins of V and $\tilde{V}_K$ have the same multiplicity.

Suppose now that we have another equidimensional algebroid variety V', of dimension r. Let $\mathfrak{O}'$; $\{x'_1, x'_2, \ldots, x'_r\}$; K' have the same meaning for V' as $\mathfrak{O}$; $\{x_1, x_2, \ldots, x_r\}$; K have for V. Let us assume that there exists a k-isomorphism $\varphi : \tilde{\mathfrak{O}}_K \xrightarrow{\sim} \tilde{\mathfrak{O}}'_{K'}$, such that $\varphi(x_i) = x'_i$, $i = 1, 2, \ldots, r$. By analogy with the case that V and V' have then equivalent singularities at their respective origins. However, such a conclusion would be meaningless since we do not have a definition of equivalent singularities in higher dimension. But we can come very near such a conclusion if we consider the case of complex-analytic varieties, in which case we can at least inquire after topological equivalence. If V is complex-analytic, the formal morphism $f : \tilde{V}_K \to V$ defines a local holomorphic map of $\tilde{V}_K$ onto V, a map which we continue to denote by f. This map is finite, i.e., for any point Q of V, near 0, the set $f^{-1} Q$ is finite. But Proposition 9.4 tells us that given any irreducible analytic subvariety W of V passing through the origin, the full inverse image $f^{-1}\{W\}$ is an irreducible subvariety W' or $\tilde{V}_K$ passing through the origin $\tilde{0}$, and that, furthermore, W and W' are bimeromorphically equivalent. In such a situation it is easy to show that f is a local homeomorphism.

Thus V and its saturation $\tilde{V}_K$ are locally homeomorphic varieties.

It follows that if there exists a k-isomorphism

(3) $$\varphi : \tilde{\mathfrak{o}}_K \xrightarrow{\sim} \tilde{\mathfrak{o}}'_{K'}$$

such that

(4) $$\varphi(x_i) = x'_i , \quad i = 1, 2, \ldots, r ,$$

i.e., if $\tilde{V}_K$ and $\tilde{V}'_{K'}$ are analytically isomorphic complex-analytic varieties (with the additional condition (4)), then V and V' are locally homeomorphic varieties.

In the case in which V and V' are hypersurfaces it is possible to obtain a stronger result, under a mild additional condition. Assume that the set of parameter $x_1, x_2, \ldots, x_r$ can be extended to a minimal base $x_1, x_2, \ldots, x_r, x_{r+1}$ of the maximal ideal $\mathfrak{m}$ of $\mathfrak{O}$; similarly for $x'_1, x'_2, \ldots, x'_r$ and $\mathfrak{O}'$. Then, if there exists k-isomorphism (3), satisfying (4), the resulting local homeomorphism between V and V' can be extended to a local homeomorphism between the ambient affine space $\mathbb{A}_{r+1}$ and $\mathbb{A}'_{r+1}$ of V and V'. Thus, in this case, the hypersurfaces V and V' are topologically equivalent as embedded varieties. In the case $r=1$ this gives, in view of theorem 8.3, the classical result that equivalent plane algebroid curves are topologically equivalent (as embedded curves; or as knotted real surfaces in real 4-space).

Using the general concept of saturation and the results stated above, we can obtain a significant partial generalization of our theory of equisingularity in codimension 1 to higher codimension. We shall now indicate this generalization. We begin by analogy with § 6, by considering an analytic family

(5) $$\mathfrak{F}: f(X_1, X_2, \ldots, X_{s+1}; t) = 0$$

of algebroid hypersurfaces in $\mathbb{A}_{s+1}$ depending on a parameter t and having the same origin $X_1 = X_2 = \ldots = X_{s+1} = 0$ (whence $f(0, 0, \ldots, 0; t)$ is identically zero; without any significant change in what follows, except for more cumbersome notations, we could have assumed that the family $\mathfrak{F}$ depends on a finite number of parameters $t_1, t_2, \ldots, t_\rho$). Here f is a formal power series in $X_1, X_2, \ldots, X_{s+1}$, t, with coefficients in our algebraically closed ground field k of characteristic zero. We make the following assumptions; f is regular if X_{x+1}, has no multiple factors in $k[[X_1, X_2, \ldots, X_{s+1}; t]]$, and if we set

(6) $$f_o(X_1, X_2, \ldots, X_{s+1}) = f(X_1, X_2, \ldots, X_{s+1} ; 0) ,$$

then f_o is not identically zero and is free from multiple factors in $k[[X_1, X_2, \ldots, X_{s+1}]]$. We denote by K the algebraic closure of the field $k((t))$, and we regard equation (5) as defining an algebroid hypersurface V_t, defined over K, and with origin at $X_1 = X_2 = \ldots = X_{s+1} = 0$; V_t is thus the general member of the family $\mathfrak{F}$. We denote by V_o the special member of $\mathfrak{F}$ corresponding to $t = 0$:

(7) $$V_o : f_o(X_1, X_2, \ldots, X_{s+1}) = 0 ,$$

where f_o is defined in (6). Thus V_o is an algebroid hypersurface, with origin at $X_1 = X_2 = \ldots = X_{s+1} = 0$ and defined over k ; but we shall regard V_o as defined also over K.

We introduce the following notations :

$\mathfrak{o}_t$ = the (complete) local ring of V_t (over K); $\mathfrak{m}_t$ -its maximal ideal;

$\mathfrak{o}$ = the (complete) local ring of V_o (over K); $\mathfrak{m}$ -its maximal ideal;

$\mathfrak{o}_o$ = the (completed) local ring of V_o (over k) ; $\mathfrak{m}_o$-its maximal ideal ;

(8) $$\mathfrak{O} = k[[x_1, x_2, \ldots, x_{s+1};t]] = k[[X_1, X_2, \ldots, X_{s+1};t]]/(f) ;$$
$\mathfrak{M}$-its maximal ideal .

Then, it is clear that $\mathfrak{o}$ =completion of $K[\mathfrak{o}_o]$ with respect to the maximal ideal $K\mathfrak{m}_o$. If we denote by $\bar{x}_1, \bar{x}_2, \ldots, \bar{x}_{s+1}$ the f-residues of $X_1, X_2, \ldots, X_{s+1}$, in $K[[X_1, X_2, \ldots, X_{s+1}]]$, by $\xi_1, \xi_2, \ldots, \xi_{s+1}$

the f_o residues of $X_1, X_2, \ldots, X_{s+1}$ in $k[[X_1, X_2, \ldots, X_{s+1}]]$ and by $x_1, x_2, \ldots, x_{s+1}$ the f-residues of $X_1, X_2, \ldots, X_{s+1}$ in $k[[X_1, X_2, \ldots$ $\ldots, X_{s+1}; t]]$, then

$$\mathfrak{m}_t = \mathfrak{O}_t . (\bar{x}_1, \bar{x}_2, \ldots, \bar{x}_{s+1}), \mathfrak{m} = \mathfrak{O} . (\xi_1, \xi_2, \ldots, \xi_{s+1}),$$
$$\mathfrak{m}_o = \mathfrak{O}_o . (\xi_1, \xi_2, \ldots, \xi_{s+1}) = \text{and } \mathfrak{M} = \mathfrak{D} . (x_1, x_2, \ldots, x_{s+1}, t) .$$

Furthermore, if $\mathfrak{p} = \mathfrak{D} . (x_1, x_2, \ldots, x_{s+1})$ is the prime ideal defined in $\mathfrak{D}$, then $\mathfrak{O}_t$ = completion of $\mathfrak{D}_{\mathfrak{p}}$ and $\bar{x}_i = h(x_i)$, where h is the canonical homomorphism of $\mathfrak{D}$ into $\mathfrak{O}_{\mathfrak{p}}$. Furthermore, $\{\bar{x}_1, \bar{x}_2, \ldots, \bar{x}_s\}$ and $\{\xi_1, \xi_2, \ldots, \xi_s\}$ are systems of parameters in their respective local rings $\mathfrak{O}_t$ and $\mathfrak{O}$, while $\{x_1, x_2, \ldots, x_s, t\}$ is a system of parameters of $\mathfrak{O}$ (in view of the regularity of f in X_{r+1}).

Remark. The canonical homomorphisms h is an injection if and only if we have $f_i(0, 0, \ldots, 0; t) = 0$ (identically) for each irreducible factor f_i of f in $k[[X_1, X_2, \ldots, X_{s+1}; t]]$.

Definition 9.5. The family $\mathfrak{F}$ is said to be equisaturated if there exists a K-isomorphism of saturated rings

$$\text{(9)} \qquad \varphi : (\tilde{\mathfrak{O}}_t)_{K((\bar{x}_1, \bar{x}_2, \ldots, \bar{x}_s))} \xrightarrow{\sim} (\tilde{\mathfrak{O}})_{K((\xi_1, \xi_2, \ldots, \xi_s))} ,$$

such that $\varphi(\bar{x}_i) = \xi_i$, $i = 1, 2, \ldots, s$.

If such an isomorphism exists, then it can be shown that the above canonical homomorphism h is an injection, and therefore , in that case, the $\bar{x}_i$ can be identified with the x_i .

A more precise way of phrasing the above definition would have been to say that $\mathfrak{F}$ is equisaturated with respect to the system of parameters $x_1, x_2, \ldots, x_s$ of $\mathfrak{O}_t$. For if $\{x'_1, x'_2, \ldots, x'_s, x'_{s+1}, t\}$

is another basis of $\mathfrak{M}$ (the parameter t being unchanged), and $\bar{x}_i'$, ξ_i' having the corresponding meaning for the rings $\mathfrak{o}_t$ and $\mathfrak{o}_o (= \mathfrak{V}/\mathfrak{V}t)$, then the existence of an isomorphism (9) does not imply the existence of a similar isomorphism φ', with $K((\bar{x}))$ and $K((\xi))$ being replaced by $K((x'))$ and $K((\xi'))$.

The following is a generalization of Theorem 6.1 (the equisingularity criterion) and is stated here without proof.

<u>Theorem 9.6.</u> $\mathfrak{F}$ <u>is an equisaturated system if and only if the</u> X_{s+1}<u>-discriminant</u> $D(X_1, X_2, \dots, X_s;t)$ <u>of</u> f <u>is of the form</u> $D_o(X_1, X_2, \dots, X_s)\ (X_1, X_2, \dots, X_s;t)$, <u>where both factors are power series (with coefficients in</u> k) <u>and where</u> $\varepsilon(X_1, X_2, \dots, X_s;t)$ <u>is a unit</u> $k[[X_1, X_2, \dots, X_s; t]]$.

We now consider, as in § 7, an algebroid hypersurface

$$(10) \qquad V : f(X_1, X_2, \dots, X_r, X_{r+1}, X_{r+2}, \dots, X_{r+s}) = 0, \qquad s \geqq 1$$

in affine $\mathbf{A}_{r+s}$, with origin $Q : X_1 = X_2 = \dots = X_{r+s} = 0$, where we assume that f has no multiple factors in $k[[X_1, X_2, \dots, X_{r+s}]]$ and the "linear" variety $W : X_r = X_{r+1} = \dots = X_{r+s} = 0$, of codimension s on V (and hence of dimension $r-1$), is contained in V. We denote by R the general point $(t_1, t_2, \dots, t_{r-1}, 0, 0, \dots, 0)$ of W, where the t's are analytically independent parameters over k, and by V_t the W-transversal section of V at R, defined over $K = k((t_1, t_2, \dots, t_{r-1}))$ by

$$(11) \qquad V_t : f(t_1, t_2, \dots, t_{r-1}; X_r, X_{r+1}, \dots, X_{r+s}) = 0 .$$

Thus, V_t is a hypersurface in affine $\mathbf{A}_{s+1}/K$. Equation (11) also defines an analytic family of hypersurfaces in $\mathbb{A}_{s+1}$, depending on r-q parameters $t_1, t_2, \ldots, t_{r-1}$. We assume that $f(0, 0, \ldots, 0; X_r, X_{r+1}, \ldots, X_{r+s})$ is not identically zero, that it has no multiple factors and that f is regular in X_{r+s} (and hence may be assumed to be a monic polynomial in X_{r+s}).

Definition 9.7. The hypersurface V is said to be equisaturated at the origin Q, along W, if the analytic family (11) is equisaturated.

The main result that we have about equisaturation (in the complex domain) is a generalization of the topological result stated at the very end of § 7, in the special of codimension 1 :

If V is equisaturated at Q along W and if V_0 denotes the W-transversal section of V defined by $f(0, 0, \ldots, 0; X_r, X_{r+1}, \ldots, X_{r+s}) = 0$, then V, as an imbedded variety in the complex affine $\mathbb{A}_{r+s}$, is topologically equivalent, locally at Q, to the direct product $V_0 \times W$ (see Zariski [9], p. 1019).

Going back to equisaturated systems and using the notations introduced earlier in this section, one derives easily from (9) the following : if $\bar{\mathfrak{O}}$ and $\bar{\mathfrak{o}}_0$ denote the integral closure of $\mathfrak{O}$ and $\mathfrak{o}_0$ respectively, and if the family $\mathfrak{F}$ is equisaturated, then $\bar{\mathfrak{O}}$ can be canonically identified with the ring $\mathfrak{o}_0[[t]]$.

Applying this result to the analytic family defined in (11) one finds that if V is equisaturated at Q along W then V and $W \times V_0$ are "bimeromorphically equivalent." However, in general, V and $W \times V_0$ will not be analytically equivalent.

In particular, Theorem 9.6 shows that in the case of codimension s = 1, equisaturation of V at Q along W implies

<u>and is implied by equisingularity of</u> V <u>at</u> Q, <u>along W.</u>

10. The Whitney-Thom Conditions

Two differentio-geometric conditions (refered to below as conditions A and B), formulated by Whitney and Thom, concern the behaviour of an algebroid variety V along an irreducible subvariety W, at a simple point of W. These conditions are equivalent, in case of codimension W=1, to our equisingularity condition (§ 7), and will no doubt play an important role in any future general theory of equisingularity (in any codimension).

For simplicity of exposition we shall restrict ourselves to <u>irreducible algebraic</u> varieties V and algebraic subvarieties W of V, in order to be in position to use, without any <u>ad hoc</u> explanations, the concepts of specialization and valuation.

However, with a few unessential modifications, everything that we state in the sequel is valid also if V is not irreducible (but equidimensional, however) and is algebroid (in particular, if V is a complex-analytic variety). The ground field k is still assumed to be algebraically closed and of characteristic zero.

Let dim V = r, dim W = ρ, and let Q be a <u>simple</u> point of W. We denote by P the general point of V and by T(V, P), T(W, Q) the tangent space of V at P and of W at Q respectively. We assume that we are also given a definite embedding of V in an affine n-space $\mathbb{A}_n$ and that Q is, say, the origin.

If R is <u>any</u> point of $\mathbb{A}_n$ and q is any positive integer

$< n$, we can represent canonically the set of linear q-space L_q in $\mathbb{A}_n$, passing through R, by the Grassmann variety $G_{n-1,q-1}$ of the linear (q-1)-spaces in the hyperplane at infinity H_∞ of the $\mathbb{A}_n$, by associating with each such L_q the intersection space $L_q \cap H_\infty$. Applying this to the case $q = r$ we consider the product variety $V\ G_{n-1,r-1}$ and the point $P^* = P \times T(V,P)$ of that variety. We can speak of the specializations of P^* (over k) and, in particular, since $Q \in V$, we can speak of the specializations of P^* over the specialization $P \xrightarrow{k} Q$. The point P^* is the general point of an irreducible subvariety V_T of $V\ G_{n-1,r-1}$, birationally equivalent to V, and the birational transformation $V_T \to V$ is everywhere regular on V^*, while its inverse in biregular outside the singular locus of V.

<u>Definition 10.1. The pair</u> $\{V, W\}$ <u>is said to satisfy condition A at Q if for any specialization</u> $P\ T(V,P) \xrightarrow{k} Q \times T_o$ <u>over</u> $P \to Q$ <u>we have</u> $T_o \supset T(W,Q)$.

In the complex-analytic domain we could have phrased this definition by considering sequences $\{P_i\}$ of <u>simple</u> points of V which converge to Q and by requiring that for <u>any</u> such sequence and for <u>any</u> accumulation element T_o of the corresponding sequence $\{T(V,P_i)\}$ we should have $T_o \supset T(W,Q)$.

To formulate condition B we consider the Grassmann variety $G_{n,1}$ of the <u>lines</u> of the projective closure $\mathbb{P}_n$ of $\mathbb{A}_n$, we consider a general point R of W such that P and R are k-independent points and we introduce the product variety $V \times W \times G_{n-1,r-1} \times G_{n,1}$. On that product variety we consider the point $P \times R \times T(V,P) \times PR$, where PR is the line joining P and R, and its specializations over the specialization $P \times R \xrightarrow{k} Q \times Q$.

<u>Definition 10.2. The pair</u> $\{V, W\}$ <u>is said to satisfy condition B at</u> Q <u>if for any specialization</u> $P \times R \times T(V, P) \times PR \xrightarrow{k} Q \times Q \times T_o \times \ell_o$ <u>(over</u> $P \times R \xrightarrow{k} Q \times Q$) <u>we have that the line</u> ℓ_o <u>is contained in the r-space</u> T_o

In the complex-analytic domain we could have phrased this definition by considering <u>pairs</u> of sequence $\{P_i\}$, $\{R_i\}$ of <u>simple</u> point P_i of V and <u>simple</u> points R_i of W, which both converge to the point Q, and by demanding that for <u>any</u> two such sequences and for any accumulation element (T_o, ℓ_o) of the corresponding sequence of pairs $\{T(V, P_i), P_iR_i\}$ (assuming $P_i \neq R_i$, for all i) we should have $\ell_o \subset T_o$.

The main reason why the Whitney-Thom conditions A and B have a claim to our attention in connection with a possible general theory of equisingularity is the following property claimed by Thom in [5] (the proof there is very sketchy and we do not claim to having it fully understood):

<u>If the pair</u> V, W <u>satisfies both conditions</u> A <u>and</u> B <u>at</u> Q <u>then,</u> V, <u>as an imbedded variety in</u> $\mathbf{A}_n$, is <u>topologically equivalent, locally at</u> Q, <u>to the direct product</u> $W \times \Gamma_{n-\rho}$, <u>where</u> $\Gamma_{n-\rho}$ <u>is a</u> W-<u>transversal section of</u> V <u>at</u> Q.

Certainly, if this property is fully proved, it would point to some sort of equisingularity of V at Q, along W, although, of course, topological triviality of along W at Q is in itself no guarantee of equisingularity.

We have a more solid foundation for a connection between the equisingularity and the Whitney-Thom conditions in the special case in which V is a hypersurface and W is of codimension 1 . We have namely the following result (Zariski [8] , Theorem 8.1) :

Theorem 10.3. *If V is a hypersurface and if W has codimension 1, then V is equisingular at Q along W if and only if the following two conditions are satisfied:*

1) *V is equimultiple at Q , along W .*
2) *The pair {V, W} satisfies conditions A and B , at Q.*

Note: In a paper in course of publication [3] , Hironaka has shown, by topological considerations, that for complex-analytic varieties, *equimultiplicity of V along W , at Q , is a consequence of conditions A and B in the most general case* (i.e. , when V is not necessarily a hypersurface and W is of any codimension) . Thus, in Theorem 10.3, condition 1) can be omitted.

The proof of Theorem 10.3 is based on the following *Jacobian criterion of equisingularity* for algebroid hypersurfaces (see Zariski [8] Theorem 5.1 and Theorem 5.2) :

Theorem 10.4. *In the notations of § 7, let $\bar{\mathfrak{O}}$ be the integral closure of the local ring $\mathfrak{O}$ (=k [[$x_1, x_2, \ldots, x_{r+1}$]]) in the total ring of quotients F of $\mathfrak{O}$, and let J be the ideal generated in $\mathfrak{O}$ by the r+1 partial derivatives $\partial f/ \partial x_j$, i=1 2, . . . , r+1 . Then Q is a singular point of dimensionality type 1 if and only if the following conditions are satisfied:*

(a) J *is a principal ideal.*
(b) Q *is a simple point of the singular locus of* V .

Furthermore, if Q is a singular point of dimensionality type 1 then the following conditions are equivalent :

(a') $x_1, x_2, \ldots, x_r$ are transversal parameters of $\mathfrak{o}$.

(b') $J = \mathfrak{o} . \partial f / \partial x_{r+1}$.

Assuming this theorem , we outline the proof of Theorem 10.3; we shall, however, restrict ourselves, for simplicity, to the case in which the algebroid hypersurface V is irreducible at Q .

In the first place we may assume, as in § 7, that $x_1, x_2, \ldots, x_r$ are transversal parameters of $\mathfrak{o}$, and that W is defined by the equations $X_r = X_{r+1} = 0$. In that case, condition 1) of Theorem 10.3 is equivalent to the following condition :

(1) $$x_{r+1}/x_r \in \bar{\mathfrak{o}} .$$

If $\bar{\mathfrak{m}}$ denotes the maximal ideal of $\bar{\mathfrak{o}}$ then it can be shown that condition A is equivalent to

(2) $$\partial x_{r+1} / \partial x_i \in \mathfrak{m} \quad i = 1, 2, \ldots, r-1$$

[compare (2) with Proposition 10.5 below], while condition B is equivalent to

(3) $$\partial x_{r+1} / \partial x_r - x_{r+1}/x_r \in \mathfrak{m}$$

[compare (3) with Proposition 10.6 below] . Now (1), (2), and (3) together imply that $\partial x_{r+1} / \partial x_i \in \bar{\mathfrak{o}}$, for $i = 1, 2, \ldots, r$, showing that $J = \bar{\mathfrak{o}} . \partial f / \partial x_{r+1}$. Thus, by the first part of Theorem 10.4, it follows that conditions 1) and 2) of Theorem 10.3 imply equisingularity of V at 0, along W. The converse follows easily from the second part of Theorem 10.4. The above relations (1) and (2) are a special case of a valuation-theoretic formulation of conditions A and B.

For illustrative purposes, we shall give here this formulation in the case in which V is an irreducible algebraic hypersurface

$$f(X_1, X_2, \ldots, X_r, X_{r+1}) = 0 ,$$

and W is the linear space $X_{\rho+1} = X_{\rho+2} = \ldots = X_{r+1} = 0$. Let $k(V) = k(x_1, x_2, \ldots, x_{r+1})$ be the function field of V.

Proposition 10.5 A necessary and sufficient condition in order that the pair $\{V, W\}$ satisfy condition A at the origin Q is that the following inequality hold true for any valuation v of $k(V)$ centered at Q :

$$\min.\{v(\partial f/\partial x_1), v(\partial f/\partial x_2), \ldots, v(\partial f/\partial x_\rho)\} > \min\{v(\partial f/\partial d_{\rho+1}), v(\partial f/\partial x_{\rho+2}), \ldots, v(\partial f/\partial x_{r+1})\} .$$

Proposition 10.6 . If $\{V, W\}$ already satisfies condition A at Q , then a necessary and sufficient condition that $\{V, W\}$ satisfy condition B at Q is that the following inequality hold true for any valuation v of $k(V)$ centered at Q :

$$v\left(\sum_{i=1}^{r+1-\rho} x_{\rho+i}\, \partial f/\partial x_{\rho+i}\right) > \min.\{v(x_{\rho+1}), v(x_{\rho+2}), \ldots, v(x_{r+1})\} + \min.\{v(\partial f/\partial x_{\rho+1}), v(\partial f/\partial x_{\rho+2}), \ldots, v(\partial f/\partial x_{r+1})\} .$$

Although conditions A and B have been formulated in terms of a given embedding of V in an affine space, it can be proved that those conditions are intrinsic local properties of V at Q , in the sense that if they hold for one embedding of V they hold for all embedding of V . In the case of condition A it is even possible to give this condition an intrinsic formulation (in which only the local rings of V and W at Q are involved) . This formulation is due to Kunz. (see [4] , unpublished) and is as follows :

Let S denote the set of all divisors of k(V) centered at Q. We set $\mathfrak{O}$ = local ring of V at Q, $\mathfrak{O}_1$ = local ring of W at Q. For any v in S we danote by R the valuation ring of v , by M the maximal ideal of R and by $\Delta = R/M$ the residue field of v (to say that v is centered at Q means therefore that $R \supset \mathfrak{O}$ and that $M \cap \mathfrak{O}$ is maximal ideal of $\mathfrak{O}$) . We denote by $D_k(\mathfrak{O})$ the differential module $\mathfrak{O}$ ' over k, and by $D_k(\mathfrak{O}_1)$ the differential module of $\mathfrak{O}_1$ over k. Since $\mathfrak{O}_1\, \mathfrak{m}_1 (\cong \mathfrak{O}/\mathfrak{m})$ is canonically a subfield of Δ , we can regard Δ canonically an $\mathfrak{O}_1$-module. We can therefore consider the $\mathfrak{O}$-module.

(4) $$R \otimes_{\mathfrak{O}} D_k(\mathfrak{O})$$

and the $\mathfrak{O}_1$ module

(5) $$\Delta \otimes_{\mathfrak{O}_1} D_k(\mathfrak{O}_1) .$$

In addition to the canonical surjection $R \to \Delta$, with kernel M , we have also a canonical surjection $\mathfrak{O} \to \mathfrak{O}$, with kernel I = prime ideal of W in $\mathfrak{O}$. This yields a canonical homomorphism $D_k(\mathfrak{O}) \to D_k(\mathfrak{O}_1)$, whose kernel is $\mathfrak{O}\, D_k I$, and therefore a canonical homomorphism of the module (4) into the module (5) :

(6) $$\sigma : R \otimes_{\mathfrak{O}} D_k(\mathfrak{O}) \to \Delta \otimes_{\mathfrak{O}_1} D_k(\mathfrak{O}_1) ,$$

whose kernel is

$$M \otimes_{\mathfrak{o}} D_k(\mathfrak{o}) + R \otimes_{\mathfrak{o}} D_k I .$$

Let $T(R, \mathfrak{o})$ be the torsion submodule of $R \otimes_{\mathfrak{o}} D_k(\mathfrak{o})$. The Kunz formulation of condition A is as follows :

Proposition 10.7. The pair $\{V, W\}$ satisfies condition A at Q if and only if for every v in S it is true that the homomorphism (6) sends $T(R, \mathfrak{o})$ into zero ; or equivalently -if and only if

$$T(R, \mathfrak{o}) \subset M \otimes_{\mathfrak{o}} D_k(\mathfrak{o}) + R \otimes_{\mathfrak{o}} D_k I .$$

Using this proposition it can be shown that $\{V, W\}$ satisfies condition A at the general point of W.

Much shorter is the proof that (V, W) satisfies also condition B at the general point of W. This is practically evident if W is a point (i.e., if dim $W = 0$). If dim $W = \rho > 0$ and R is a general point of W/k, then the proof can be easily reduced to the case $\rho = 0$ by a ground field extension $k \rightarrow k^* =$ algebraic closure of the field $k(R)$.

We shall now conclude with some remarks concerning the global aspects of conditions A and B, say, on affine varieties $V \subset \mathbb{A}_n$.

We consider again the irreducible subvariety V_T of $V \times G_{n-1, r-1}$ defined by its general point $P \times T(V, P)$. Similarly, let W_T be the irreducible subvariety of $W \times G_{n-1, \rho -1}$ defined by its general point $R \times T(W, R)$. Here P and R are k-independent general points of V and W. We set

$$H = V_T \cap (W \times G_{n-1, r-1})$$

and we consider on $W \times G_{n-1, r-1} \times G_{n-1, \rho-1}$ the variety

$$(7) \qquad \varphi = (H \times G_{n-1, \rho-1}) \cap (W_T \times G_{n-1, r-I})$$

Let Z_o denote the algebraic subvariety of $G_{n-1, r-1} \times G_{n-1, \rho-1}$ consisting of those pairs $(L_{r-1}, L_{\rho-1})$ of lienar spaces which satisfy the incidence condition $L_{\rho-1} \subset L_{r-1}$, and let

$$Z = W \times Z_o.$$

Both Z and φ are subvarieties of the product variety $W \times G_{n-1, r-1} \times G_{n-1, \rho-1}$. It is easily seen that if Q is a simple point of W, then $\{V, W\}$ satisfies condition A at Q if and only if

$$(8) \qquad \varphi \cap (Q \times G_{n-1, r-1} \times G_{n-1, \rho-1}) \subset Z .$$

By the definition (7) of φ, we have a projection map $\psi : \varphi \rightarrow W$ (everywhere regular on φ). By (8), the set of simple points Q of W such that $\{V, W\}$ satisfies condition A at Q · is the complement, on W, of the set $S_o \cup \psi\{\varphi - (\varphi \cap Z)\}$, where S_o is the singular locus of W. Since $\varphi - (\varphi \cap Z$ is an open subset of φ, it follows that the ψ-transform of this set is a contructible subset of W.

We have thus shown that the set of simple points of W at which $\{V, W\}$ satisfies condition A is constructible. Since we know (as was pointed out above) that this set contains the general point of W, it follows that the set contains a non-empty open subset of W.

Using similar considerations it can be shown that also the set of simple points of W where $\{V, W\}$ satisfies condition B is constructible (and therefore contains a non-empty open subset of W).

It may very well happen that the set of points of W where condition A is satisfied is not open (or-equivalently - that the set of points of W where condition A is not satisfied is not an algebraic subvariety of W; we know, however, that this set is contained in some proper algebraic subvariety of W). The following is an example of this possibility:

Let V be the hypersurface $X_4^2 - (X_1X_2+X_3)X_3^2 = 0$ in $\mathbb{A}_4$ and let W be plane $X_3 = X_4 = 0$. Thus W is the entire singular locus of V, and is a double plane of V. The set of points of W where condition A is not satisfied is the union of the two coordinate axes $X_2 = X_3 = X_4 = 0$ and $X_1 = X_3 = X_4 = 0$, from which, however, the origin $X_1 = X_2 = X_3 = X_4 = 0$ has been removed (at the origin, $\{V, W\}$ does satisfy condition A).

We know of no example in which the set of points of W at which both conditinns A and B are satisfied is not an open subset of W. However, in the special case in which V is a hypersurface and W has codimension 1 and is singular for V, the set of points in question is the set of points of W which are not of dimensionality type 1, and this set is algebraic (see § 7, result 4)).

O. Zariski

Bibliography

[1] S. Abhyankar, On the ramification of algebraic functions, American American Journal of Mathematics, v. 77 (1955).

[2] ________, Inversion and invariance of characteristic pairs, American Journal of Mathematics, v. 89 (1967).

[3] H. Hironaka, Normal cones in analytic Whitney stratifications, Institutut des Hautes Etudes scientifiques (1969), Publications Mathématiques, No. 36 (volume dedicated to Oscar Zariski).

[4] Ernst Kunz, Uber gewisse singuläre Punkte algebraischer Manningfaltigkeiten (1969, in course of publication in Crelle's Journal).

[5] R. Thom, Ensembles et Morphismes stratifiés, Bulletin of the American Mathematical Society, v. 75, No. 2 (1969).

[6] H. Whitney, Local properties of analytic varieties, Differential and Combinatorial Topology, A symposium in honor of Marston Morse Princeton, 1965).

[7] O. Zariski, Studies in Equisingularity I. Equivalent singularities of plane algebroid curves, American Journal of Mathematics, v. 87 (1965).

[8] ________, Studies in Equisingularity II. Equisingularity in codimension 1 (and characteristic zero). American Journal of Mathematics, v. 87 (1965).

[9] ________, Studies in Equisingularity III. Saturation of local rings and equisingularity, American Journal of Mathematics, v. 90 (1968).

[10] O. Zariski and P. Samuel, Commutative Algebra, v. 1 and v. 2, (D. Van Nostrand Co., Princeton, USA, 1958 and 1960).

[11] Frédéric Pham et Bernard Teissier, Fractions Lipschitziennes d'une algèbre analytique complexe et saturation de Zariski Centre de Mathématique de l'Ecole Polytechnique, n. M 170669 (June, 1969).

Editoriale Grafica - Roma

GPSR Compliance

The European Union's (EU) General Product Safety Regulation (GPSR) is a set of rules that requires consumer products to be safe and our obligations to ensure this.

If you have any concerns about our products, you can contact us on ProductSafety@springernature.com

In case Publisher is established outside the EU, the EU authorized representative is:

Springer Nature Customer Service Center GmbH
Europaplatz 3
69115 Heidelberg, Germany

Batch number: 09490862

Printed by Printforce, the Netherlands